M000223043

This book is a "Dragons Claw Press" Publication
2519 5th Ave.
Belle Fourche, South Dakota
www.paleoadventures.com
To order wholesale copies for retail distribution:
Telephone orders: 605-892-2634
Email: stein151@earthlink.net

Proudly published in the United States of America

First edition:
Copyright ©2009 Walter W. Stein
All Rights Reserved

No part of this book may be reproduced or utilized in any form or by any means,
electronic or mechanical, including photocopying or recording, nor may it be stored in
a retrieval system, transmitted, or otherwise copied for public or private use, without
the expressed written permission of the author.

ISBN #0-9716206-1-X

ILLUSTRATIONS:
Copyright ©2009 Walter and Heather Stein
All Rights Reserved.

Original concept, prototypes and cartoon character design: Heather Stein
Art direction, layout design, text and captions: Walter Stein
Illustrated by Greg Fisk
Cover Illustration: Greg Fisk
Cover Design and layout: Walter Stein

A special thanks to the talented Greg Fisk, who enabled our ideas to really come to
life!

BACKCOVER:
Photos courtesy of the Munn Family 2007
Thanks for making a face for radio appear marginally acceptable! LOL!

To the Piedmont City Library,
Thanks for inviting me
to speak at the library.
I hope you had a good time.
Best Wishes and happy hunting,

THE TOP 256 RULES OF PALEONTOLOGY

... PRACTICAL ADVICE FOR FOSSIL TECHNICIANS

BY:

WALTER W. STEIN

ILLUSTRATED BY GREG FISK, WALTER W. AND HEATHER STEIN

©2009 WALTER W. STEIN- ALL RIGHTS RESERVED
A DRAGON'S CLAW PRESS PUBLICATION

A DEDICATION

THIS BOOK IS DEDICATED TO MY LOVELY WIFE AND MY TWO WONDERFUL SONS. THANKS FOR ALLOWING ME TIME TO EXPLORE MY ECCENTRIC PURSUITS...

***Special thanks to: Valerie Nakamura for her inspiration and assistance with editing, Greg Fisk for his wonderful illustrations, Mike Triebold for giving me the tools to make a difference and also to all those paleo-hungry kids out there with an open mind to dream. Your smiles and innocence inspire me to do the same...

PHOTOGRAPH OF THE AUTHOR ON A TYLOSAURUS ("SOPHIE") DIG OUTSIDE OF WACO, TEXAS IN 2004. IT WAS UNFORTUNATELY DURING THE MIDDLE OF RAINY SEASON!

TABLE OF CONTENTS

I. INTRODUCTION

During the summer of 2003, I had the wonderful opportunity to be both the curator for the Rocky Mountain Dinosaur Resource Center (RMDRC) and the Field Collections Manager for Triebold Paleontology. We had just finished constructing the RMDRC, a fantastic new museum in Colorado, and I had found myself trying to accomplish multiple tasks with about ten different staff members who had little or no direct field and laboratory experience. Anyone who has ever been in such a position knows that as middle management, it's your responsibility to make sure you get the boss's goals achieved as quickly and efficiently as possible. You have your own jobs that must be done and you have to make sure your staff is still doing theirs. As middle management, you spend half of your time arguing with the boss and the other half yelling at the staff. As a result, and since I couldn't be everywhere at any given moment, I decided to put together a little inter-office memo I called, the *Top 256 Rules of Paleontology,* for our newbie paleo-techs.

Now, the *Top 256* started off as kind of a joke. Some of the crew were more difficult than others and rather than repeat myself over and over again, I wrote down a few of the techniques that I wanted them to follow. I tried to make it funny and light hearted, so as not to be perceived as a fascist dictator who did nothing but run around all day and yell at people. The original 256 list really consisted of less than 40 guidelines and then skipped ahead to rule 256 which was of course "Do not glue yourself to the bone!". It's a long story, but we had one intern who did this on an hourly basis, hence the rule. Over time, the list grew to include field work and lab work techniques and though there was still confusion periodically (the intern still continued to glue himself to everything in sight no matter what we tried), people basically began to use those techniques on a regular basis. Efficiency, productivity, communication and quality began to slowly improve.

In 2005, my wife and I decided that it was time to take our paleo show on the road, and start our own company called PaleoAdventures (www.paleoadventures.com). The bread and butter of the company was and still is, doing educational dinosaur dig-site tours on private land in South Dakota, Montana and Wyoming. As I would do these tours, I found myself periodically, almost accidentally, incorporating some of the

Top 256 rules into my tour speeches. Guests loved it and most picked up the techniques fairly quickly. The bones were safer, the guests got a few laughs and the tours ended without anyone falling off a cliff or bitten by a rattlesnake.

By 2007, our company had grown to the point that the tours were really going well. I hired an assistant of mine, Valerie Nakamura, who was one of the bright paleo-technicians I worked with at the RMDRC. She was so good, that we made her education director at the museum. When she also decided to leave the museum and went back to teaching we remained in contact and remained friends. During the summer of 2007, we had enough operating cash, to hire her on, part-time, to help with our field tours. Inevitably, the *Top 256* came up again. As we joked about the rules one sunny July day, it dawned on me… if the *Top 256* helped to train paleo-amateurs into disciplined careful paleo-technicians, then perhaps other people might find some use for it as well. At least they might get a laugh or two. So, in August of 2007 I began writing this guide, filling in all of the top 256 rules I thought young students or people thinking about a career in paleontology might need to know. Many people have asked, "Why 256 Rules?" The answer is; "I have no Earthly idea!" It was simply the number that popped into my twisted head. I could have easily made it 100 or 365 or 1000; there is certainly enough information to do that. "256" just seemed to fit. As I continued to write however, the book took on a shape similar to my first book, "So You Want to Dig Dinosaurs?" and hence the *Top 256* has become a companion to that (ahem!) masterpiece.

Hopefully, if you've purchased this book you are either well on your way towards a career in paleontology as a student or are actively involved in volunteer or paid positions somewhere within its fold. I sincerely wish you well in your journey. I am not so conceited as to think that this book can replace traditional educational routes nor substitute for hands-on training by someone far more important than I. Hopefully though, readers will be able to find a few of the tips and advice contained in its pages useful and as stated before at least humorous. Best wishes to you on your paleontological quests and hopefully we will meet somewhere in the badlands surrounded by thousands of pristine, scientifically important skeletons. Happy hunting!

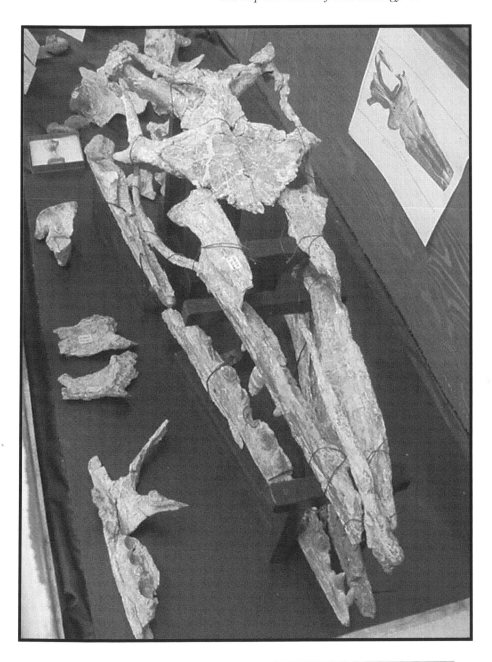

PHOTOGRAPH OF "MARY THE MOSASAUR"- A *MOSASAURUS MISSOURIENSIS* SKULL AND PARTIAL SKELETON DISCOVERED BY THE AUTHOR IN 2005 IN BUTTE COUNTY, SD.

II. THE TEN... NO WAIT! TWELVE COMMANDMENTS

Okay, okay, okay... so I'm not Moses coming down from the top of Mount Sinai with some burning tablets. The pages should not be smoking and lighting will probably not strike you down should you, in the course of your life's passion, break one of these rules- point taken... However, these top twelve are probably the most important things you should take heed of when pursuing your path of paleontology. These are things that apply to introductory students as well as advanced PhDs. Hopefully, they will guide you, comfort you, give you an edge and most of all keep your ego grounded. Failure to follow these rules will not kill you with a lighting bolt, but in a worse case scenario, failure to follow these rules might kill your career or torpedo it before it can even get started.

1. Paleontology is a "Career" not a "Job":

If you are looking for a simple, consistent, routine, 9-5 job, turn back now, because you are definitely going in the wrong direction. There is no such thing as a "job" in paleontology. Do not expect to work 8 hour days with 1 hour lunch breaks. Do not expect time clocks. Do not expect your day to end when a whistle blows. Do not expect to have "normal vacations", sick days, health benefits or union lawyers. Do not expect to become super rich. Do not expect to be so comfortable in your knowledge base that you can stop learning and survive. Do not expect to sit all day long in an air-conditioned cubicle.

Paleontology is a life's passion. It is a quest of biblical proportions- seeking the answers to some of life's greatest mysteries. One does paleontology not for a pay check, but for the pursuit of knowledge. If you expect to do paleontology, expect to work 10-14 hour days under a blazing sun, in a busy prep lab or buried in a research library. Expect studying til the wee hours of the night, (3-5 days/week) just to stay current on the latest discoveries. Expect to be doing ten different projects (from field work to teaching) on any given day. Expect each and every day of your paleontological career to be unique, different and challenging.

2. **Never Give Up:** Throughout your career, you will encounter friends, family, co-workers, teachers, neighbors, strangers and all manner of humanity who will encourage you to throw in the towel and get a "real job". Many will not understand your strange desire to scratch around in the dirt all day long. Family will become annoyed as you drag them from one natural history museum to the next. Teachers will tell you that there are no jobs left in paleontology. Professors will tell you that you do not have what it takes to become a paleontologist. Parents will tell you that there is no money in paleontology. The media will tell you that there is no point to studying paleontology further as we know everything there is to know. If YOU know that studying fossils and helping unlock the mysteries of life on Earth is your life's passion, DO NOT LISTEN TO ANY OF THEM. You have only failed if and when you decide to give up.

I cannot tell you how many people I have met who have said…"man, I've always wanted to do that ever since I was a little kid, but…_____" [Fill excuse in the blank]. At some point, someone stepped in front of them and said don't go any further and they unfortunately listened. For example, at a fossil show in Colorado, I met one gentleman who listened to the nay-sayers when he was a young man. He wasted 40 years of his life working a boring job he hated, only because it "paid well". Now he is approaching retirement, chasing a meager pension, too afraid to start over and after looking back, seeing many regrets. You do not want to be that person forty years from now. Life is far too short to spend it doing something that does not inspire you. If dinosaurs and woolly mammoths, ammonites and trilobites inspire you to think, do, create, imagine, wonder, study and work hard then find a way to fit them into your life.

Yes, there are very few jobs in paleontology. Yes, it is very difficult to find and maintain a career in this discipline. Most positions do not pay well. Many are on a volunteer basis. Most require an advanced degree. You may need to make many personal and financial sacrifices to pursue it. You will need to be creative, pig-headed, stubborn, dedicated and unwilling to give up your dream. No one ever said life was "easy", why should the study of ancient life be any easier. If you can accept those things… you have a shot to make it.

3. Cherish the Moments of Discovery:

There are few things in this life more incredible than the sense of discovery. When you have toiled all day long for weeks at a time and you uncover something that no human being has ever laid eyes on; something that sunlight hasn't shone upon in over 65 million years, that is an amazing feeling. When after months of research you discover something surprising about one of those ancient specimens that no previous researcher has ever thought of, that is job satisfaction. When you have been hiking up and down drainages for weeks, covering miles upon miles exhausted, and you see the first few bones of a new skeleton emerging from the rock (and they don't resemble anything you or anyone else has ever seen), that is joy.

Paleontology can best be described as this: hours upon hours, months upon months, years upon years of intense, slow, patient, challenging physical and mental work punctuated by short periods of incredible glee. Cherish those moments, for that is the reason paleontologists do what they do.

4. Patience is Essential: Patience is unfortunately not

something that can be taught or learned. You do not find it in a text book. Either you have it or you don't. Those that have inherited patience have the potential to do well in paleo. Those who don't, unless directed creatively, do badly. Whether you are in the field, the lab, the research room or the museum, you will need patience.

The records of ancient life did not wait in the ground 180 million years + for you to come along one day and carelessly #$@#?$*&!!! them up! You have a responsibility to the science and to the fossils to treat each natural treasure as if they were the Earth's last. In order to do this, you must exude superhuman patience in the face of whatever frustration may come along. It is like the race between the tortoise and the hare. The hare moves quickly, but recklessly. He is a danger to the fossil, himself and to others around him. He bounces about appearing to get things done, but in the end, achieves little and usually does more harm than good. The tortoise on the other hand, moves slowly and cautiously, examining everything. He is always laying out a carefully crafted strategy so that any obstacles that he encounters are anticipated and defeated. To the casual observer, the tortoise may not seem to be doing much at all, but over time, he moves mountains.

Imagine sitting, hunched over a preparation table, in a brightly lit laboratory, working on a single bone- removing one grain of sand at a time. The process can take months, sometimes even years. I knew a preparator that spent the bulk of almost two years, working on a two-meter-long Apatosaurus sacrum, encased in a dense limestone concretion. Each day, all day, he slowly scratched, chipped and chiseled off small chunks of matrix (surrounding rock) surrounding the bone. Some days were mind-numbingly frustrating for that preparator. In the end, however, the tortoise won, and a beautiful, perfectly prepared sacrum that scientists can now study until the end of civilization, was the result.

The game of paleontology is not for those with "attention deficit disorder" or for those who become bored at the slightest of whim. Paleontology is best performed by individuals who live in what I call deep time. People who live in deep time are those who, in a chess match think and see five moves ahead of their next move. They waste little time thinking about what will be best for them or their discoveries in the short term, but see a bigger picture and place more emphasis on the long term. Long term goals, long term time commitments, long term rewards. It is a race for those who think before they react, observe before they advance, look before they leap and learn before they know.

5. Document Everything: The best paleontologists in

the world write down every observation they encounter. Even the most seemingly useless fact, may turn out to have importance. If not to you, perhaps to one of your colleagues. If not to them now, it might have importance to someone studying your finds 50 years in the future. For this reason, good paleontologists make a concerted effort to document everything they see in the field, the lab and the research room.

Documentation involves recording the geology, stratigraphy, taphonomy and provenance of a fossil dig site. It includes written and photographic/video documentation. It includes detailed measurements and observations about the dig site and the fossils contained within it. You will record your information collected in the field in a field journal (see rule #93). You will keep a running list of specimens recovered from the site (see rule #94), outlining who found them, their orientations, their preservation, and where they are currently stored. You will map (rule #100 and 101), your dig site showing the relative and exact positions of the elements. You will draw cross sections and stratigraphic columns (see rule #99). You will do a detailed analysis of the rocks at the dig site. You will record the history of the discovery and the collection. You will map out the location of the dig site and provide a detailed site description. You will note the geochemistry of the rocks surrounding the fossil and the nature of the preservation and fossilization. You will collect microfossils and rock samples. You will record your observations during fossil preparation, take photographs of the process and make drawings and sketches of the elements. You will document everything! Failure to collect the contextual data along with the specimen is not an option.

6. Reputation is Very Important: As stated

before, there are very few people who are actively engaged in paleontological pursuits. Jobs are scarce and competition for esteemed positions is fierce. If you do something so bone-headed, that it makes headlines you can believe me that the word about your foolishness will get around. Once it does, you can forget about ever having a chance for any position of importance ever again. Not only does word spread fast in this small community, it has lasting power because paleontologists think in the long term not the short. If you are known as the "guy who carelessly smashed a giant vertebra with a hammer and chisel right in front of a *National Geographic* cameraman", then guess what… everyone will know it sooner or later. If you have been proposing that dinosaurs went extinct due to "extreme flatulence", then pretty much, you will be known as the "dinosaur fart guy" for most of your career. Unless of course, you come up with something even more bizarre.

The point is simple: Your reputation, whether good or bad, will follow you for your entire career. It will help or hinder you when looking for employment. It will help or hinder you when publishing or working in conjunction with your peers. Most paleontologists are known by their work not their face. Only the pretty ones make it to TV, so whatever hypotheses you propose will define you and follow you good or bad.

7. **Ethics, Ethics, Ethics:** Having a good reputation involves more than just what you propose and where you propose it. Sometimes your reputation revolves more around how you have proposed your theories and how you have conducted yourself. Most professional paleontological organizations have a code of ethical standards that their members are expected to follow. Both the Society of Vertebrate Paleontology (SVP) and the Association of Applied Paleontological Sciences (AAPS) have such a code and it is routinely enforced. These codes include things which protect members from stealing each other's research, or publishing data or images without giving credit. They encourage members to be careful about ethical collection practices and to not trespass on private land without prior permission. They suggest that important specimens should be

deposited in a public repository. They emphasize which ethics and practices are considered good and which are considered bad.

If you do decide to break these rules, you are often at risk of being permanently expelled from these professional organizations. Sometimes, these unethical practices will follow you for the rest of your career. You do not want to be known as "the guy who embezzled money from the not-for-profit foundation", or "the girl who knowingly stole a skeleton from BLM land", or "you know, that guy who steals all his graduate students research", or worse yet, "the guy who flirts with all his graduate students (and then steals their research)". Those people don't work too much in this field any more. Many of them have gone through terrible legal battles and some even went to jail. You do not want to be one of those people. You do not want to be known as "the drunk", "the letch", the "trespasser", the "thief", the "crook" or the "jerk", for those brands could potentially follow you for the rest of your career. Therefore, be ethical in all that you do.

8. **Follow the Rules of Logic:** Every good scientist knows that the rules of logic dictate good quality research and fair and honest scientific debate. Since the dawn of civilization, mankind has known that in any debate, there are valid arguments and invalid arguments. Arguments for a particular hypothesis are valid, IF those arguments follow the rules of logic. Over time, abundant supporting evidence and careful scrutiny, many of those valid hypotheses may become theories. Arguments for a particular hypothesis that do not follow the rules of logic, are invalid, and often breed misinformation, mis-instruction, contempt, anger and confusion. Too often many paleontologists today, including some of the most esteemed doctorates, use arguments for their hypothesis that can easily be dismissed as being illogical. Knowing and recognizing fallacies of logic will aid you in interpreting the arguments and conclusions of other scientists and will help you to write better research papers of your own. Once you know, recognize and understand them, you will be less likely to fall prey to illogical reasoning. The following lists

some of the more important fallacies (out of dozens) in logical reasoning you should be aware of:

A. **Ad homonym**- This fallacy states that an author claims are invalid because of some irrelevant fact about the author. For example: *The author, Kevin, suggests that birds evolved from dinosaurs, but Kevin is not a "nice guy", so birds could not have evolved from dinosaurs.* The conclusion may or may not be true but not as a result of the premise.

B. **Call to authority**- This fallacy is the reverse of Ad homonym. It says that what an author claims MUST be valid, because the author is a well known expert. For example: *The author Kevin, has a PhD and over 20 years of experience. He suggests that birds evolved from dinosaurs so they must have.* The above statement may or may not be true, but not as a result of the premise.

C. **Non-sequitur**- (That which does not follow) A non-sequitur is a claim followed by a statement which does not have anything to do with the original claim. The conclusion is not supported by the premises. For example: *Eagles have wings, therefore birds evolved from dinosaurs.*

D. **False or weak analogy**- This fallacy uses an analogy as evidence for a conclusion, but the comparatives between the two are not necessarily equal. For example: *Dinosaurs are like birds, therefore all dinosaurs could fly.*

E. **Argumentum ad populum**- This fallacy calls on the reader to suspend disbelief because a large group of others believe a certain way. For example: *90% of scientists polled now agree that birds evolved from dinosaurs, therefore it must be true.*

F. **Circular argument-** (aka. Begging the question)- This common fallacy uses a premise to prove a conclusion which then uses the conclusion to prove the premise. It is an attempt to prove two statements reciprocally from each other. For example: *Dinosaurs went extinct gradually because they do not speciate as much in the late Cretaceous. We know they were not speciating as much, because they were slowly going extinct.*

G. **False dichotomy-** This fallacious argument presents two separate premises which from which to choose, as the only possibilities for an outcome. For example: *Either you believe that the dinosaurs died out gradually or you don't believe in extinction.*

H. **Red herring-** A red herring occurs when an author presents a statement that is completely unrelated to the original argument. It is designed to confuse the reader and throw them off track. For example: *Birds are obviously not the descendants of dinosaurs... as we all know dinosaurs went extinct.*

I. **Straw man-** A straw man is an argument against a particular premise that first distorts the original argument (or position) and then attacks the new, distorted, unrelated position. This gives the impression that the original argument or position was false. For example*: Dr. Jones: "I believe we should allow some commercial collectors to collect fossils from public lands. It is better for these specimens to be in private hands then for them to erode away to dust.* Dr. Bonehead responds: *"The unrestricted mining of fossils from public lands would be a travesty of science. It is better for the fossils to remain where they are, than to allow every rock hound with a hammer and chisel to rip* *them from the ground."* Dr. Bonehead, in this example, has used the fallacy of Straw man in his argument because Dr. Jones never argued for anything along the lines of "unrestricted mining" or "ripping things from the ground".

J. **Argument from ignorance** (argumentum ad ignorantiam)- This fallacy of logic argues that a premise must be true in the absence of anything which proves it false or that a premise must be false in the absence of anything that proves it true. For example: *"The Permian extinction could not have been caused by an impact. Where is the crater? There isn't any crater."*

K. **Scare tactics** (ad baculum) – This fallacy of logic uses bullying and intimidation to cause a fallacious argument to seem logical even when it isn't. For example: *"If we allow commercial groups to collect fossils on public land, they could potentially mine out all of our precious, non-renewable, fossil treasures."*

L. **Biased sampling-** This occurs when a researcher omits certain data (intentionally or unintentionally) leading to a biased data set. For example: *A study of the last ten million years of the Cretaceous confirms that there are only 12 valid genera (8 potential others were omitted and ignored) in the Late Cretaceous, confirming that dinosaurs were on a slow and steady decline.*

M. **Fallacy of composition-** This occurs when the characters of part of a group are assumed to be representative of the characters for the entire group they are composed of. For example: *A commercial paleontologist I once knew was corrupt, so an organization composed of many commercial paleontologists must be equally corrupt.*

N. **Fallacy of division-** This occurs when the characters of a group are assumed to also be the characters of an individual within that group. For example: *The "x" academic association is the most respected in the world, therefore since Joe Bonehead is a member of that group, he must be one of the most respected paleontologists in the world.*

O. **Hasty generalization-** A hasty generalization is when an author jumps to conclusions with limited evidence. For example: *We've recovered three Dromaeosaurs with feather impressions from Liaoning Province, therefore ALL Dromaeosaurs had feathers.*

P. **Post hoc-** A post hoc fallacy occurs when an author assumes that a particular event that happened before a second event, in the absence of additional information, caused the second event: For example: *Everyone knows that sea levels were falling before the end of the Cretaceous. This must have triggered the extinction of the dinosaurs.*

Q. **Slippery slope-** A slippery slope fallacy involves a chain of events where one thing inevitably leads to the next without question. For example: *If we allow commercial paleontologists to collect on public land, they will collect and sell every fossil they find. We will not be able to control it. Soon there will be no fossils for professional scientists to collect because the commercial groups will have mined out each and every one of them, until there are none left* (boy that's efficiency for ya).

9. Know Your Prey: Research, Research, Research:

If you were a wildlife biologist and you wanted to do research on the Nile crocodile, do you think it would be a wise move to jump on a plane, fly to Africa, immediately jump in a boat and begin chasing Nile crocodiles without having ANY knowledge whatsoever about Nile crocodiles? No! Of course not! It would be suicide to try and study a large dangerous animal on its home turf, without knowing as much as possible about them before you began doing field observation. You would want to study the animal from a distance first. You would want to learn from the experiences of others. In this example, you might learn what the animal looked like so you could properly identify it in the field. You might want to learn what its known habits were and what habitats it preferred so you could successfully track it and study it. You would learn what it ate and how it ate it and if it was overly aggressive or shy and passive. You would need to know as much as possible about it, BEFORE you went looking for it. The same applies to fossil beasts (without the threat of being crunched, spun about and stored to soften as in the above example!). Before we go looking for fossil beasts, it's a good idea to know a little something about them. The key to knowing your prey is research.

Research is one of the most important things a paleontologist must do during their career. If you are studying a particular type of organism or a particular process, then you must read and study as much information about that organism or process as possible, before you try to publish your own observations. You should visit and study known specimens up close and in private. You will need to know their anatomy like the back of your own hand. You should study as many research papers as possible about the organism and consult your peers who have worked with them in the past. Once you know your prey, inside and out, then and only then, should you ever attempt to comment on it.

10. Good Scientists Remain Skeptical: Every

good scientist is a skeptic by their very nature. Don't believe me? Good, then you're being skeptical and you've got a shot. Remember this rule when reading this or any other book. The great philosopher David Hume is famous for saying; "All I know is that I do not know". In other words, the only thing that we can truly prove to be 100% correct is that we simply "do not know". This applies directly to paleontology, as the objects of our study have been dead and gone for often hundreds of millions of years. Without the invention of a time machine, studies involving behavior, growth rates, physiology, and the like can never be proven beyond a shadow of doubt and must always remain in question. New data, specimens or observations might be discovered that will render those hypotheses invalid. If we simply accept information, data, observation or hypotheses as cold, hard facts, we fail to allow ourselves to accept the possibility that they are wrong. When we do this, we are no longer scientists but drones. We become "proponents" or "advocates". We no longer test things to prove them to be the case, we simply assume them to be accurate. You probably know what happens when you "assume" too much. Therefore, when you read a scientific paper, or study a persons position on a particular topic (any topic), first ask yourself what their assumptions are. Then test those assumptions to determine for yourself whether or not you consider them to be valid and logical. Once that is achieved you can then go through the paper, skeptically, asking yourself all manner of questions including; was the data collected correctly(?), was the methodology sound (?), could there be underlying motivations behind the conclusions (?), is the data subjective or inductive(?), could the results have been contaminated in any way (?), are the observations different than my own (?), are their any untestable leaps of faith (?), etc. etc. etc. Once you have done this, and assuming the paper or opinion or whatever has held up to that kind of scrutiny, then, and only then should you even begin to accept it. Even after all that, you must still remain skeptical, for if you turn your back on new information, new data, new specimens that disprove the previous paper, dismissing them out of hand, you will have lost your way from true science.

11. You're not Satan, But Others May Disagree: As a paleontologist, you will often be working with or working on, controversial theories that seek to answer questions surrounding the origin of life (evolution, etc) and the processes of death (extinction, etc.). Inevitably, you will encounter individuals who are also doing this, not from a scientific angle, but a religious one. For those individuals, these issues were traditionally answered by "faith" and not necessarily by "facts". In the past, these questions were dissected using what little knowledge of the universe those individuals had at that time and a faith that the stories passed down from one generation to the next, were based in both truth and "divine" inspiration. We should not, in hindsight, fault our ancestors for it.

For most of civilization's history, the questions surrounding "life" and "death" were the sole jurisdiction of religion. Science was the interloper. As science began to evolve within the human psyche, however, (testing things, analyzing things, questioning things) there was an inevitable and unfortunate conflict between the two sides, even though they were both trying to answer many of the same questions. The trouble was, and still is, that each side operates under two distinctly different methods, speaking in two distinctly different languages. One group chooses to "believe" by "faith" the other chooses to "prove" by "scientific method". Religion wonders why science can never produce "concrete" truths about the universe and science wonders how people can live by untestable leaps of faith. Though both sides have their great strengths and great weaknesses, trying to reconcile the methods and the answers is like trying to compare apples to oranges. It can not be done. Most of the planet's population sees the world in a religious light, and scientists, though we may disagree with religions conclusions, need to understand that only through patience, tolerance, education and acceptance will we be able to "reconcile". For some, religion based upon faith will always be their method of choice and that's okay. Just accept that! Evangelizing from either side is a waste of time and energy.

As a future or current paleontologist, you are bound to encounter people who still believe the Earth is only 4,000 years old. You will meet people who will argue that the theory of evolution is bunk. On rare occasions, you may even encounter some, who believe that all paleontologists are Satan worshipping atheists, bent on destroying their family's way of life, by digging up the bones of dragons, which Satan had buried to confuse mankind (WHEW!). Some of these people, you probably know, respect and even admire. Some might even be members of your own family.

Often this conflict between science and religion can make it very difficult for paleontologists to gain access to prime, fossil collecting lands on certain private ranches. Earlier last spring (2008), I was attempting to gain access to just such a ranch when I encountered a female rancher, with just such a belief system. After a quick introduction, the conversation (from her end) turned decidedly nasty. She basically told me that she was "sick of all those paleontologists spreading lies about that evolution nonsense". "I don't want you out there. The bones are better off left in the ground". Now, she knew there were dinosaur remains on the ranch... she choose to ignore them. Unfortunately, some folks prefer to bury the evidence rather than expose it to the light of day. That's just how it is sometimes.

If you are going to be involved in paleontology, irregardless of your own personal religious views, expect to hear such criticisms or arguments against you and your chosen profession from time to time. If it gets too overbearing just try to ignore it. If you do choose with your "free will" to be drawn into the fray, please try to do so with mutual respect and courtesy and please try real hard not to make fools of the rest of us.

(SOAP BOX RANT- Just in case you are wondering and in order to give full disclosure, I am actually an agnostic- not an atheist. Unlike many scientists today, I recognize that atheism is an indefensible, illogical, non-scientific position. To believe that any single individual can know, for an absolute fact,

without skepticism, further inquiry, or experimentation, that there is NO god/supreme being/prime mover/grand architect (whatever)is well beyond the scope of what science can do and outside the limited capability of human logic, reason and understanding. This absoluteness in belief puts the atheist on the exact indefensible plane of faith as the theist. I argue that it is impossible to prove whether or not there is or is not a supreme being, and thus a glorious waste of time for a scientist to debate it in a scientific context (i.e. don't try to use science to disprove God). If you would like to take off your science hat and put on your philosophers robe then by all means do so. My problem is that many scientists attempt to blend their science and their religion and use it to alter public policy. This is crossing a boundary that, I would argue, is dangerous to both science and religion and more importantly public policy. I believe in religious freedom and that our religion can and should remain separate, personal and private. For my religious friends, I would also argue that the natural processes of evolution and extinction work just fine in the presence of a divine being; that it is equally plausible that they are merely the tangible expression of that beings handiwork. Science simply must stop at a certain point or it runs the risk of becoming as guilty of dogmatic close-mindedness as many traditional religious followers. It is for these reasons that I accept the POSSIBILITY of a higher power AND the POSSIBILITY that none exists. It's a <u>literal</u> translation of Genesis in the Bible that I can't really agree with based upon the overwhelming evidence to the contrary. Hell, a day in the life of a God could, in theory, be billions of years for all we know!?   At least I think I know that?   I do like those Ten Commandments though! Whew! Glad that bit is over!)

12. Do Not Be Afraid- The Fossils Can Smell Fear: Well okay, the fossils can't

really smell anything- they're dead. You, on the other hand, are very much alive and can cause great harm to them and the field of paleontology if you are afraid of doing anything with them. This applies to all aspects of paleo. For example, if, in the field, you are afraid you will damage something, you will piddle about, digging a few grains at a time and never, ever find anything important no matter how many months you dig. If you are working on a stratigraphic column and you make an observation that is contrary to the opinions of one of your partners, but you are too afraid to contradict them, it's quite possible that the mistake may be repeated by others for decades. If you are told to use a certain lab technique that you know would be inappropriate to use, but don't say anything and do it anyway, that fossil might be destroyed in the process. If you believe a superior scientist is incorrect in a hypothesis, but are too afraid to voice an opinion, the science could well be set back a hundred years. Fear is a terrible thing and has no place in the fields of science or paleontology. Fear is what holds back advancement, both personal and professional. Imagine what the world of paleontology would look like if Charles Darwin did not overcome his fear and publish, "On the Origin of Species". Imagine that Bob Bakker was too afraid to write "The Dinosaur Heresies"…we'd still have sluggish cold blooded evolutionary failures gracing the pages of children's books. Imagine if Walter and Louie Alvarez were too afraid to test for iridium in the K-T boundary clays or Mary Higby Switzer too afraid to look for original dino-proteins. It is through these fearless leaps of faith, through experimentation, that the best techniques, the best theories and the best decisions are made. Of course there are going to be failures, mistakes, and incorrect paths along the way, but even in these, science finds use and ways to sort it out in the end. You must be fearless, but not reckless. You must be confident and courageous. The best paleontologists are.

III. RULES FOR FIELDWORK

Paleontologists often spend all day long, under a blazing sun, staring at the dirt upon the ground. No matter how much certain paleontologists try to make paleontology into a quantitative, mathematical or theoretical science, it will always predominantly be a field science based on subjective observation. You will, as paleontologists, spend much of your time in the field, traveling to remote, often wild places, looking for clues; clues that will hopefully help to answer the complex geological and biological questions scientists have been contemplating for years. You will be collecting raw data, the very foundation of paleontology. You will be on the front lines of the discipline. If you do not get field experience, you will miss a very important component in the understanding of ancient life and your education will be incomplete. The following list of rules is designed to help novice or inexperienced amateurs to quickly begin working with fossil resources in a safe and careful manner.

A. GENERAL FIELD RULES:

13. You are not God (or Superman)-Accept that: The first rule of field work is that you are not God. You are not all-knowing, all-seeing, omnipotent or omniscient. You are not Superman. You do not have X-ray vision. Without the aid of some geophysical tool, you can not see into the rock or the dirt beneath your very feet. You do not know what is there. Below you could be a perfectly preserved Tyrannosaurus rex skull or... absolutely nothing at all. Knowing and accepting this, should give you two powerful tools. The first is "temperance"- knowing that if you swing a pickaxe into the rock too hard, in the wrong place or the wrong way, you might put it straight into a priceless specimen,

destroying it forever. The second thing is "grounding"- the knowledge that, no matter how cautious you are, you will make mistakes. Every non-god out there does.

14. You Will Break Stuff: Sorry, I know it's hard to accept, but after all you're not God, right? Since you cannot possibly see into the earth, no matter how cautious you may be, you will break stuff. There isn't a field or lab person out there who hasn't. If you encounter a paleontologist who says they have never accidentally damaged a specimen in the field they are either 1) lying or 2) have spent no time in the field. Accepting this rule is hard, but necessary; otherwise the fear of causing any damage will cripple you. Whereas rule #13 slows you down and makes you think before you dig, rule #14 softens the blow of the inevitable.

15. If You Do Break It, Do So in a Way that it is Repairable and Recoverable: This rule is the key to proper excavation and will hopefully help you to find a pace and rhythm that is appropriate for you and your site. If you excavate too fast you most assuredly will break something badly, causing irreparable damage. If you dig too slowly, you will not get anything accomplished. With limited time and money for field work, neither of the above is a suitable outcome. We must, after all, be practical. Therefore, if we can accept that breaking things is inevitable, but not preferred, we must dig fast enough to achieve the goals of our excavation, but slow enough that if something is hit, it can be repaired (in the field or the lab) and recovered. Since, each specimen is different, each site is going to be different and each field worker is different, finding the "right" pace is difficult. There are a lot of variables that will give you a justification for either slowing down or

speeding up. For example, if you are working in a dense bone bed, where fossils are found in great quantities, then you need to take an extra slow pace. If you are working in a site where the invertebrate fossils are spaced apart, but are generally pretty solid, you might take a faster pace. If you are working with very common fossils where damage to one is not the end of the world, speed up. If you are working on a one-of-a-kind piece, that you dare not even scratch, slow down to a crawl. Use common sense and be practical.

The speed in which you dig will help you to limit any field damage, but it is only half of the equation. The other half is the force that you employ with each stroke. The best technique is to use uniform and even pressure irregardless of what tool you are employing. Violently, unevenly, stabbing at the rock with a pickaxe, trowel, or blade is probably not going to be a smart move in a known fossiliferous layer. Specimens and specimen parts are sure to fly everywhere. They will be irreparable and irrecoverable. If you use slow, even pressure with the appropriate tool, should you hit something, you almost always realize it and can back out, causing minimal damage if any.

In the terrible, but inevitable event that you do damage something in the field, don't try to hide it or ignore it. Take a deep breath, collect the pieces, label them and hope that you can put Humpty Dumpty back together again in the lab. If you were following rule #15, you will be able to repair it and make it close to its original condition. If you were not following rule #15, have fun trying to explain your box of powdered fossils

16. Each Dig Site Has it's Own Set of Rules: Each and every dig site has its own set of rules. Those rules depend on a wide variety of things including the nature of the matrix, the time constraints, the safety and security of the team, the type of fossils you are

collecting, the importance of the fossils you are collecting and many other factors. Those variables will determine the speed in which the site is worked, the tools used and the skill levels needed to do it. In reading the following rules regarding field work, please keep that in mind. Many of them are not hard and fast, set in stone rules. It is very important to remain flexible and adaptable when working with fossil dig sites.

B. EXPLORATION AND LOGISTICS:

17. Logistics are More Important than You Think: Logistics involves coming up with a complete plan for how you are going to approach any field season. In most cases, you don't just wake up in the morning, jump in the truck and start banging on rocks. It's usually more complicated than that. You do not want to find yourself on a dig and realize that you didn't bring enough plaster or burlap, you forgot the glue or you forgot to get all of your permits taken care of. No. In order to be successful, you must have a plan and you must be able to execute that plan. You must know and understand what your objectives are and be ready to meet them. You must have all of the proper excavation and safety equipment. You must have a team of support staff qualified and capable enough to assist you in meeting your goals. You must have maps to the field locations, acquire transport where necessary, acquire permits or signed contracts and know where you can and can't explore. You have to establish and develop budgets and timetables, contact locals or guides for assistance and have everything ready to go and in place long before you hop in the truck and head down the road. Conducting an exploration program

without addressing these issues is foolish and will lead to mistakes and poor judgments in the field.

18. Choose a Field Crew Wisely: Who
you decide to take with you on your collecting trips is also very important. You do not want to be in the middle of Mongolia and discover that you have teammates who hate each other, can't speak the local tongue or have dubious reputations. If you do not pick your support staff carefully, expect everyone to be miserable as conflicts are bound to occur. When that happens, the excavation and documentation of the sites will suffer.

19. Get It in Writing: If you intend to work on
private land with private land owners, be sure to ensure that they know exactly what it is you are doing out there. Prior to contacting them for permission, speak with an attorney and have that lawyer draw up a formal, legal contract, outlining all the expectations that you have and all the safeguards necessary for both you AND the land owner. This contract needs to be as detailed as possible so no misunderstandings, leading to extended, embarrassing legal disputes, occur in the future. Do not write your agreement on a cocktail napkin! Don't accept a mere handshake- it will not hold up in court. Don't consider a nod and a wink sufficient. Do not compromise on this, otherwise it will come back to haunt you, sooner or later. Many fossils have a high commercial value (a certain $8.3 million dollar T. rex immediately comes to mind) and when people start adding up dollar signs and start getting greedy, they often do dumb things. If you have everything clearly spelled out in a legal contract there is less of a chance that someone is going to break their word and do something silly. It is for both your protection as well as the land owner's protection. Do not set foot on private land without this contract!

#18- There was just one spot left on this season's field crew... Unfortunately, none of the over 400 applicants were qualified!

20. Visit The Courthouse- Buy the Platt Maps:

Once you have contacted a private land owner and had them sign a legal land use contract, the next step would be to visit the local county courthouse and research the official legal property boundaries of the land in question. Don't take the land owner's directions/descriptions as law; double check! Generally, you will need to go to the tax assessor's office where they register titles and land surveys. In some cases, you can buy from the courthouse a "Plat Map", which already has the legal property boundaries and land owner's names drawn onto it. If you get that opportunity, you need to purchase it. If not, you will need to spend some time transferring the legal property boundaries onto USGS Topographic Maps.

21. Transfer Property Boundaries to a 24,000 Scale USGS Topographic Map:

USGS topographic maps can be purchased online or over the phone from many sources. They generally cost a mere $3.00-$6.00 per map. A large ranch or search area might require three to four topographic maps to completely encompass its boundaries. So, you are not looking at a terrible expense. The USGS topo maps will be invaluable to you for finding your way or plotting potential dig sites on either private or public land. Once you have the maps, I recommend that you fold them in half several times, then tape the folds (on the back of the map) to prevent them from fraying. Then, transfer, using a marker pen and colored pencils, the property boundaries of your private or public land where you have permission to explore, onto each map. When exploring, consult the maps frequently.

#22- EVERYONE DREADS THE "SCARLET LETTER T"!

22. No Trespassing! The absolute biggest no-no for any field paleontologist or any amateur rock-hound is to intentionally walk onto land that you do not own and start beating on rocks. Anyone who pursues paleontological specimens in this fashion should be shot and hung up by their toenails. These criminals give the entire field a bad name and their actions have led to many private and public lands being declared off limits for even legitimate investigators. If you are reading this book and you intend to trespass, please pick the book up and smack

yourself in the face until common sense enters your brain or you die of exhaustion. Thank you.

23. Be Professional- Be Responsible:

You need to follow the ethical codes discussed in rule #7, and remember to act accordingly. Remember you need to be professional and courteous when dealing with private landowners, townspeople and fellow paleontologists. Don't go running around the local establishments trash talking your peers, the locals or (duh!) your land owners. That's a good way to make everyone in paleontology look bad and unprofessional. Also, everyone knows that access to private lands is limited, but there are plenty of sites to go around. Fossiliferous rock formations make up a large part of the continental United States. If you get shut out of one area go search for another. Don't threaten, bully or con a private land owner into turning over their infamous dig site to your university/company/etc... "or else". This attitude is the number one reason why many private land owners now deny everyone access (and... duh(?)... if no one is allowed to collect, then whatever is out there will slowly be destroyed). Don't try to manipulate land owners who already have someone working on their ranch, to gain access for yourself. This is very unprofessional and completely unnecessary in most cases. Of course, it is your responsibility to maintain good relationships with your land owners, so if you get the boot due to your incompetence and someone else moves in to take your place, you have no one but yourself to blame.

24. It's Their Land Not Yours- Treat Private Land-Owners with Kindness and Respect:

As a paleontologist, irregardless of whether you are an academic or a commercial type, remember that you are not only representing yourself to land owners, but you are also indirectly representing the rest of us too. If you work on private land, be friendly, honest, open and courteous to land owners, locals and other paleontologists who may also be working in the area. Remember that if you are working on someone else's property, they pay the taxes and therefore make the rules, not you.

25. Fence Lines Do Not Necessarily Equal Property Boundaries: In the

Western United States, large scale, private ranches have a way of running into one another. Often times land owners have built fences without the benefit of an official survey. Sometimes, fences are built due to geographical necessity, conforming to the terrain and not the title. In some cases, ranchers lease land from other ranchers and often there is no marker in the field to separate the two. All of these factors mean that fences are not always a reliable indicator of legal property boundaries. Fence lines could be off by several dozen feet in bad cases or not even present at all, in worst cases. There are many examples of people who thought they had found a specimen on private land, where they had permission, only later to realize (to their horror) that the location of the boundary did not correspond to the fences and they were unfortunately trespassing. Whereas honest, legitimate accidents sometime happen out in the middle of nowhere, be very careful when exploring. Do not assume the fences are accurate.

#25- DUE TO SOME UNKNOWN, BIZARRE LAW OF NATURE, THE BEST SPECIMENS TENDED TO DIE ALONG FUTURE PROPERTY BOUNDARIES!

26. Public Lands? You Better Have a Permit:

In the western United States, public lands make up a large portion of the surface area. Private and public lands are often intermingled and "checker-boarded" with one another. So much so, a chunk of public land could be isolated right in the middle of a private ranch with no markers whatsoever. The BLM and the National Forrest Service both sell maps which outline the boundaries of the land under their stewardship. Contacting local BLM agents or forest rangers and getting to know them is always a good thing to do. So long as you are open and honest about what your intentions are, most local agents will be as well. Know where the private land ends and the public land begins. State lands and even some county lands have agents that can tell you the laws and regulations regarding fossil collecting there. As a general rule however, it is illegal to remove any vertebrate fossil remains from National Park, National Forest, National Monument, BLM, BOC, or other federal lands without a permit, an approved collection plan, an impact study and a nationally recognized, federally approved, public repository backing your efforts. I recommend contacting each agency to find out their policies regarding invertebrate collection. In nearly all cases though, you will either need to have an advanced degree or someone with an advanced degree on the permit application in order to get approval to collect on public land. If you do wish to work on public lands, follow the stipulations on that permit to the letter or expect to pay heavy fines, potential jail time or a loss of reputation that you will not be able to recover from.

27. Know and Follow the Local Collecting Laws:

States, certain counties and certain towns also have particular laws regarding fossil collection that you need to be aware of. If there is any doubt of the legal nature of your exploration or excavation work, I highly recommend that you contact the appropriate governmental agencies to know your rights and their requirements. Sometimes contacting the nearest university or museum will point you in the right direction.

28. Never Look Where You Do Not Want To Dig: One of my favorite rules, taught to

me by the great Michael Triebold, is to never look where you don't want to dig. It is a wonderful, frustrating, waste of time to do otherwise. If you spend all day exploring a three hundred foot vertical cliff face and find a skeleton sticking out of the bottom, what have you achieved? The specimen is unlikely to be collected without removing the entire mountainside! If you find it one hundred feet below the peak, how exactly are you going to excavate it? How will you get it out? Unfortunately, some specimens are impossible to access, so don't look where this is going to be a problem. Knowing you have a nice specimen in one of these locations will keep you up at night, racking your brain trying to figure out how it could be done, but it doesn't often lead to a positive conclusion.

29. When in Doubt, Stay on Established Roads: This rule applies to public

and private lands. Going "off road" in many public areas will land you a heavy fine if you get caught and can cause some environmental problems if the area needs reclamation thanks to your carelessness. Going "off road" on private land without prior permission from the land owner can often cause you to be kicked off the land permanently. To a rancher, grass is gold. If you destroy a pristine pasture by running back and forth over it with your vehicle twenty times expect to encounter an angry rancher when you come out for the day.

30. Close Those Darn Gates! Surprisingly, this is one of the most common problems with access I hear from local ranchers. Hunters, bird watchers, government land surveyors, and yes, paleontologists, are well known for forgetting to close barb wire fence gates as they drive carelessly across private land. This absolutely drives rancher's nuts. I mean, have you ever tried chasing a stray cow in the middle of August?! Be polite.

31. GPS Your Sites and Make Sure You Are Where You Think You Are! Let's say you have found a prospective site and you are ready to begin digging. Hold on and wait just one minute. Are you absolutely positive you are where you think you are? Take 10 minutes, GPS it, plot it and verify your location. Don't waste two weeks of time and money only to find the embarrassment of trying to explain how you went off course to an angry neighbor, your land owner or worse, a federal agent.

32. It's Always Best to Explore with a Partner: Often times, your paleontological pursuits will take you to some seriously out of the way neighborhoods. For this reason it is always best to hunt with a partner. Hopefully, a partner you can trust. If you explore by yourself and accidentally fall down a cliff, break a leg, twist an ankle or something else worse along those lines, your partner can help get you to safety.

33. Aerial Reconnaissance without an Airplane- Google Earth:

Exploration for fossils can be aided by using aerial photography to find potential outcrops and map geologic rock units. Google Earth, a wonderful, free, relatively new computer program can provide you with an aerial image of your target area without you ever leaving home. By comparing aerial imagery with topographic maps you can cut your exploration time requirements in half. Of course, you will still need to be on the ground, on site, to truly analyze the fossil potential of an area, but aerials definitely help get you close. You can download the Google Earth program from their website. Companion sites also offer additional downloads that you can add-on, which may help with exploration or research. These include things such as; volcanic or earthquake data, geospatial data and surveying add-ons, etc, etc. etc.

34. Have You Considered Under-Explored Rock Formations?

Many paleontological field workers focus their attention on rock units that have been previously worked on with great success. Rock units are not created equally. Some have more fossils than others. Rock formations such as; the Niobrara Chalk, The Pierre Shale, the Hell Creek Formation, The White River Fm. or the Morrison Formation are world renowned for the important fossils that have been plucked from their depths. It is important that explorers continue to probe those formations as more important fossils will no doubt be recovered from them until the end of civilization. With that said, however, it is also equally important to personally investigate other rock units that are not always known for their fossil assemblages. In many cases these rock units, the under-explored ones, can help paleontologists fill in many of the gaps in the fossil record. Don't immediately discount a formation simply because the literature tells us that it is devoid. There are plenty of examples of important fossils being found from these locations. The Cedar Mountain Formation, for instance, was not widely explored until the late 1980s and 1990s. Now, it has revealed a large suite of dinosaurs and other vertebrates that continues to grow in number and significance.

35. Think Light-Weight, Mobile, Fast and Utilitarian:
Too many field workers come ready to explore, weighed down with pounds and pounds of unnecessary gear and contraptions that no human being should be burdened with in 100 degree temperatures. When exploring, wear loose fitting clothes with lots of pockets. You do not need to fill each and every pocket with stuff! You will need flagging, collecting baggies, marker pen, a multi-tool like a *Gerber* or *Leatherman*, full water bottle, cell phone or walkie-talkie, brush, knife and shovel or walking stick. That's it. Leave the rest of the gear back with the truck or back at base camp. Only when you find something of significance should you bring the rest of your fancy gear.

This rule also applies to the expedition as a whole, not just exploration. I once had the pleasure of working with a paleo-tech that made going anywhere a challenge. Whenever we went on a fossil hunting expedition, this gentleman brought everything and the kitchen sink. He had piles of camping gear, three coolers, boxes upon boxes of food, three suitcase full of clothes, extra tools, extra GPS, dirt bike, pedometer, emergency weather band radio, etc. etc. etc. To everyone's astonishment, he once literally used a bobcat to lift his giant box of gear into the back of the field truck! Not only did the entire staff want to strangle this bonehead, but he left absolutely no room for the fossils we were pursuing. Needless to say, most of the gear "accidentally" rolled off the back of the truck before we left.

36. Dirt Merchants in a Hi Tech World- Geophysical Tools: Exploring

for fossils is often done, just as it has been done, for the last two centuries; a human being hiking up and down drainages looking for that trail of bone or field of broken shell. It will probably remain the preferred method for the next two hundred years or more. With that said, however, do not close your mind to the possibility of using other geophysical tools that might aid you in your explorations. Geophysical tools have been used with varying degrees of success on various paleo-projects. Things such as Geiger counters have been used to find large, uranium-rich dinosaur bones in Utah. Ground penetrating radar has helped scientists over potential bone beds trace and locate large bones buried underneath several feet of solid bedrock. Magnetometers and other tools also, may have some limited merit in specific cases. Though there is no single tool that makes the job of exploration a no-brainer, be open minded to experimenting with and learning about these potentially useful tools.

37. "Is this Important?" Don't Remove Potentially Important Elements From a Fossil Locality Without First Marking the Site. Many novices,

when they first encounter what appears to be an important find, immediately pick it up from the ground and excitedly start running back to the "expert" they hope can identify it. In the process they forget where they found it! On one such occasion we had a friend out collecting with us. She

#36- Dr. Bonehead's new invention was working perfectly... until he ran into something unexpected.

was very eager and had a great eye for finding specimens. Unfortunately, she was a hair too eager. Upon the discovery of a sun-bleached, medium sized, theropod dinosaur phalanx (toe bone), she rushed back through the badlands and showed the specimen to me. Recognizing the bone to be from a small Tyrannosaur, probably a *Nannotyrannus*, I got excited and asked her to take me to where she had found it. We hunted that drainage for two days straight, but we were never able to be sure where exactly it had come from. The phalanx was probably just an isolated specimen, but we could have been walking just below a fully articulated, one of a kind, *Nannotyrannus* skeleton and never found it. Uggghhh! I don't want to think about that possibility… let's move on….

38. When Exploring, Remember to Flag Potential Sites: See the example in rule #37 for the reason. There are numerous stories of a similar nature with equally frustrating results. If it looks promising, mark the site. You can always come back to it later.

39. Time is Valuable- Stay on Target:
Remember, you only have so much time and so much money for each exploration project. If your goal is to map out the K-T boundary, don't waste time piddling around in a microfossil bone bed nowhere near the K-T Boundary. If your goal is to find a specimen of an early marine whale, then don't waste time picking up seashells. It is very easy to get distracted when you love fossils as much as we do. If something does catch your eye, mark it on your map or flag it and come back to the spot if and when there is time. If you can't do it later, tell someone else who might be interested in it and have them investigate. It could be significant or it could be a waste of time. Either way,

address your first project first, before getting mired in another.

40. Your Job is To Save Those Fossils- Not Ignore them! Your job as a

paleontologist is to collect important specimens that will hopefully help unlock the mysteries of ancient life. There are many scientific studies, such as faunal diversity projects, paleo-environmental projects, predator-prey ratio studies, taxonomic studies, ontogenic, and sexual dimorphism projects that all require large samples of individual specimens to even be close to being accurate. The only way to do this is to collect lots and lots of specimens. You may have simply found "just another *Triceratops*" specimen, or "just another duck-bill dinosaur", or just another "sauropod", or another "mammoth", and individually each may not make headlines or cause *National Geographic* to call in the middle of the night, but collectively, they have value. If you do find something, even if it doesn't seem too important at the time or doesn't meet your project goals, do not entirely forget about it. Come back to it and collect it later. Or, notify someone else who might find it interesting and help them to get access. Leaving the specimen to sit out there in the weather to rot is unconscionable. Failure to collect these specimens that you or others might not consider "significant" is a waste of our precious natural resources. Anyone who steps over a mass death assemblage of hadrosaurs to collect a theropod and then permanently ignores the hadrosaurs needs to seriously re-examine why they got into this field in the first place.

41. Don't Vacuum The Badlands!

Whereas rule #40 tells us we need to save fossils from weathering and erosion, some fossil fragments are simply better left in place. It is not a good idea to pick up every little fragment and scrap of bone or every little fossil you discover. Even though erosion works fast in the badlands, removal of every scrap will obscure the horizons from which the fossils are eroding out. Leave the scraps in place, so that future explorations in the area, by yourself or other workers in the future may re-investigate those horizons for more important things at a later date.

42. No Paleontological Prairie Dogs! I

have encountered several old localities where paleontologists and paleo-wannabees have dug what they call "test pits"; giant holes in the earth, dug vertically into a prospective site. Often times this method of exploration digs straight through otherwise solid and important fossils destroying them or severely damaging them. If you are not going to approach a dig site with care and determination to finish the site, then do not even bother to dig these holes! At one such locality in South Dakota, at least four groups from both public museums and private companies had done such a thing. Their debris piles were full of discarded chunks of mostly destroyed fossil dinosaur bones. It is only common sense, that if you have an outcrop that stretches over ¼ mile, where weathered bone fragments litter the surface that should you dig into the hillside, you are bound to find more. You do not need to dig a "test pit" to find that out. Common sense should tell you that. We are not gophers or prairie dogs. If you aren't going to excavate seriously, then leave it for someone in the future who will.

43. Fossils Are COMMON- Skeletons are Rare:

One myth, frequently advanced by many in our field, is that "fossils are a rare commodity". We hear this argument again and again from those on the far left hand side of the spectrum- academic paleontologists who are trying to build a case against the private ownership of vertebrate or even invertebrate remains. These scientists often use this straw-man tactic to try and push through unnecessary and divisive fossil collecting legislation, legislation in my opinion which would be a detriment to the field. They use it to try and gain public sympathy, funding or support for their museums, research or pet projects. They use it to gain collecting monopolies in some areas. They have used it to manipulate land owners into donating valuable collecting sites or specimens with little to no compensation or recognition for those land owners. In some cases they have even used this argument to force governmental agencies into supporting efforts to confiscate private lands and the fossils contained within them out right!

While it IS true that complete and partial skeletons of CERTAIN vertebrates are rare, and these should be placed in a public repository for responsible, open study, there are many fossils that are quite common and should be available to the public for the enjoyment and education of all.

Fossils occur frequently in many sedimentary rock units all across the world. These rock units erode over time and their hidden fossil treasures likewise spill out across the deserts, badlands, hills, mountains, farmer's fields, construction sites, road outcrops and river systems in many countries. There are areas of Utah and Wyoming, for example, where fragments of sauropods and even complete skeletons are popping out all over the place. There are parts of Kansas and Canada, Argentina to Africa, Chile to China, where common specimens of vertebrates are now a dime a dozen. There are places in

Tennessee and Texas, India and Indiana, South Dakota and South Africa where invertebrate fossils literally, litter the landscape. The fossiliferous rock units that contain these wonderful pieces of ancient history dip under ground, re-surface and go back down again. These rock units are often hundreds to thousands of feet thick and cover millions of square miles. We could not completely "mine them out", in a million lifetimes. In these rock units, there is the potential for fossils of all sorts- many of them, very common.

For example, I recently returned from a fun little collecting trip along the Peace River in Florida. My licensed guide and I braved the poisonous snakes and ten foot alligators, wading out into the river with our simple shovels and screens. Within minutes we had found what we were seeking. Each and every shovelful of sediment we dug along the bottom of that river contained dozens of vertebrate fossils. There were shark teeth by the hundreds, crocodile, horse and bison teeth by the dozens and tons (literally tons) of dugong and whale bone fragments littering the bottom of that river channel! We were not "raping and pillaging the natural world" by doing this! We were not "destroying the fossil record"!

Science is losing nothing if these types of fossils wind up in private hands. From a practical standpoint, museums simply can not store or curate these specimens and most would have no desire to do so. From a scientific standpoint, there isn't anything new that could be gained if they did. A single dugong rib bone fragment can not tell us anything new about the behavior or systematics of dugongs. A single shark tooth is not going to tell us much about the life and times of ancient sharks. The same applies for other vertebrate chips and chunks, isolated bones, teeth and poorly preserved specimens which are frequently found (in the right locations) in great abundance. It does absolutely no harm to science for these to be in private collections.

In the hands of an eight year old child however, these scraps and unwanted-uncollected pieces have a tremendous scientific benefit- that little dugong bone or shark tooth, lovingly brought to school for "show and tell", might just inspire a classroom full of kids to think of something more than the latest video game for a few minutes. Perhaps, they might encourage a future paleontologist to take their first steps toward their chosen career path. I would fathom that most current paleontologists, even the ones who are now opposed to private ownership, got their start in a similar manner.

Triceratops, *Edmontosaurus*, *Mosasaurus*, *Apatosaurus*, *Camarasaurus*, *Oreodonts*, fossil fish, shark teeth, fossil horses, camels, mammoths, etc, etc, etc, are all very common in the right areas. Collecting them upon discovery protects them from the natural processes of weathering and erosion. I, for one, would rather see an oreodont on a mantle piece in someone's private home, rather then in a pile of pieces scattered across some barren stretch of lonely badlands, wasting away to dust!

Let me stress the following, however, so that we are perfectly clear: Important, unusual, unique, special, SCIENTIFICALLY SIGNIFICANT fossils, including complete skeletons, ARE rare and should be collected by experienced professionals and preserved and curated in a professional academic institution! Every attempt should be made to get them into one. Unforgiving laws, hateful rhetoric and academic monopoly are, however, no way to save our natural fossil treasures. If we truly wish to do the right thing and preserve our fossil record, it can only be achieved through common sense ideas, practical laws and honest public education.

44. Finding Fossils is Easy- Knowing What to do with them is Hard:

As stated in Rule #43, fossils are commonly found in rock formations all over the world. Invertebrate remains can be found in great abundance. Vertebrates too, in localized spots at least, can be found in large quantities. There is not, as many would have you believe, a deficit of fossils or fossil localities. Ten year old kids with no prior experience have found important fossils. Ranchers, farmers, and construction workers with no formal training frequently discover things they suspect might be important. Finding fossils is not hard. Knowing what to do with them once you do find them is what becomes difficult. If you think you have found something that might be a fossil, don't use it as a doorstop or toss it in the rock garden. Don't box it up and shove it in the attic. Don't re-bury it and hope no one saw you hit it. Make significant use of it. If you are not qualified to excavate a specimen or it is beyond the scope of your skills, contact someone whom you trust, who can help. Excavate your fossils correctly... then, get the important ones into a good, respectable home where they will be cared for and studied.

45. Follow the Bone Trail: When a fossil skeleton becomes exposed to the elements it immediately begins to weather and erode. The rate that it weathers is dependant on many factors including the degree and nature of the fossilization, the type of bone, the size and density of the bone, the tectonic forces and fracturing that it was exposed to during its burial, the chemistry of the surrounding soil and rock, the hardness and density of the surrounding rock and the climate the bone is exposed to once exposed. Smaller, more delicate bones break down faster than larger better preserved ones. Specimens encased in harder matrix tend to be better preserved and weather slower than ones in softer, loosely consolidated sediments with high moisture content. Regardless of the rate, over time, the specimen breaks down into smaller and smaller fragments. These begin to get carried downhill from its horizon of origin leaving a trail that the

exploration paleontologist can follow. Depending upon the number and size of the bones in the bone bed, and the time the horizon has been exposed, the paleontologist, can determine how wide the debris field is and how easily it can be traced. Once a field worker discovers a trail of weathered bone they must trace those remains uphill until the horizon from where the weathered remains originated is discovered. Sometimes, depending upon the steepness, amount of vegetation and vertical extent of the outcrop, finding where the pieces came from is impossible. Other times, however, the trail will lead you straight to an excellent find.

46. Collect __All__ of the Float: Weathered
chunks of fossil bone are called "float", because they have "floated" or eroded downhill along the least resistant path. Most of the time these chunks of weathered fossil bone have been sitting out for so long that they are not repairable. In some cases, careful preparators can put the fragments back together again like one would a jigsaw puzzle. Since you do not know what can be repaired and re-assembled and what can't, while in the field, it is important to pick up every last fragment found below any complete or partial fossil skeleton. Too often, paleontologists in the past ignored much of the weathered remains, collecting only the larger chunks or smaller more complete bones and fragments. Sometimes the float was simply left or not thoroughly investigated.

A perfect example of why thorough float trail recovery is essential to good field work, is a specimen of a polycotylid plesiosaur skeleton I was lucky enough to discover, during the summer of 2007. Upon discovery, we diligently picked up every scrap of weathered bone we could find on the surface. Then, we began excavating the in-situ bones from the bone bearing horizon. It was only later, after a month of walking up and down the hill below the excavation, and a few good rain storms that we realized we missed quite a bit of the float. In fact, we missed lots of float. As we re-investigated the sediment beneath the dig we began finding complete bones and repairable fragments that had eroded out dozens of years ago, washed downhill and had become reburied under 3 to 24 inches of sediment. To date, over 50 complete bones and several boxes of repairable fragments have been recovered from this float trail. Many of these were over 25 feet downhill from the bone bearing horizon. It would have been very easy for me to ignore the trail and focus only on the in-situ elements. Had I done that, 10 % of that specimen would have been lost and forgotten. Imagine how many other specimens collected over the last hundred years and now residing in public and private collections all over the world, are incomplete solely because their discoverers did not collect the re-buried weathered debris.

47. If You Don't Know Where You've Been, How the Hell Will You Know Where You Are Going? Each

day after every exploration run, it is important to note where you explored, in both your journal and in some cases on your topographic maps. It's often very difficult to remember where you have looked and where you haven't after several years of hunting on one ranch or one tract of land. If you record which gullies, outcrops and drainages you have searched, you will have a better understanding, over the years, where you need to explore next. Sometimes, you may even want to keep track of which pastures have been hunted so that you can leave them "fallow" for a few years. Letting an area erode for a few years allows for the creation of new bone trails that you can then trace to your newly eroding out, potential skeletons. This rule also applies to excavation (see rule #63).

C. Excavation:

48. Use the Right Tool For the Right

Task: Paleontologists use all manner of small, medium and large tools during the excavation process. Each tool has a specific function that when used correctly is magical. If used incorrectly, a nightmare. Each tool has an upside and a downside and you should know the difference. Like a surgeon, the field paleontologist must carefully choose the correct tool or otherwise risk the patient. The surgeon would not want to use a bone saw if they only needed a fine scalpel. The same applies to the excavator. Don't use a rock hammer if you only need a flat dental probe. Don't use a jack hammer, if a half-hour of shoveling will do the trick. Many workers mistakenly try to find the tool that will move the most rock in the shortest amount of time. This can often lead to mistakes and damage to the very fossils you are trying to protect. You wouldn't use a shotgun to kill a fly (though come to think about it… that might be fun), so don't use a D-9 to scour randomly in a known bone bed. A list of some excavation tools and their preferred uses follows:

A. Dental picks, pins and probes- These light weight, easy to wield tools come in a variety of shapes and sizes. They are best suited for fine delicate work directly over crushed or highly fragmented bones. They work well in the slow removal of shale or mudstone matrix or the removal of excess glue and dirt between large fractures. Best in helping to untangle closely spaced, more delicate fossils. Their drawback is that they are exceptionally slow and remove very little material at a time. They do not work well in sandstones or loosely consolidated rocks.

B. X-acto knives or scalpels- X-acto knives work great in just about all rock types except for highly cemented sandstones. The fine tip can be used in the more delicate removal of matrix, whereas the curved blade works to slice, probe, scrape or lift soft sediments around both heavy and delicate bones with ease. The blade can also be used to cut annoying roots or to re-break fractures so they can be cleaned out and reset. They are a great multi-purpose tool with few drawbacks.

C. Brushes- Paleontologists routinely use brushes of every shape and size to remove dust and loose debris from the surface of a bone bearing horizon. I have found that the best brushes for the task are NOT the most expensive, soft bristle, Moroccan camel hair ones that cost an arm and a leg, but the cheaper, short bristled, hard bristled "chip" brushes you can find at most hardware stores. If you need to do extreme detailed prep in the field occasionally old tooth brushes also come in handy.

D. Pocket knives- Larger knives tend to work well around larger, better preserved specimens that are not tangled with other fossils. They work best in mudstones or conglomerates, where their heavier blade strength holds up better than an X-acto. The drawback is that they are not usually best for precision or delicate work.

E. Ice Picks- Ice picks work best for probing into mud or clay-rich rocks like siltstones, mudstones, claystones, and shale. It is generally difficult to use them in any precision work and they are better employed in finding the maximum extent of a fossil. If torque is necessary to remove large chunks of matrix, the ice pick is going to be the least invasive of the larger tools.

F. Trowels- A flat bladed or angle bladed trowel works best in loosely consolidated sandstones, siltstones and shales, working with the grain and lamina. Trowels do not work well with smaller, more fragile fossils as the blade can easily cut fragile fossils in half. Trowels are also important for squaring up quarry walls for cross section analysis and are often employed when trying to shave matrix clean in the preparation of applying plaster jackets. .

G. Small shovels- I have personally not had much luck using smaller garden variety hand shovels though others swear by them. They are best employed in sandstones around heavier fossils.

H. Hand rakes- These are sometimes used for combing float trails looking for fragments or for removing small debris piles.

I. Air-scribes- Air-scribes are small to medium sized precision instruments that look and act like miniature jackhammers. They have a vibrating tip that slowly chips away at stubborn matrix. Though they are predominantly used in the lab, they work well in the field if you can keep them clean, dry and oiled. They can be used when you are up against extremely hard sandstones or concretions where the goal is to isolate and separate, with precision, certain elements. They are great for surgically separating closely spaced bones that you do not wish to jacket in one block. Air-scribes come in a variety of different sizes and types ranging from small bodied, delicate "microjacks" with fine needle tips, mid-sized "Aero", or "Chicago pneumatic" scribes designed for removing larger chips in denser rock and the larger "Super Jacks" and "Mighty Jacks" designed to remove very large chips in highly cemented rocks. The scribe you will choose depends upon the type of matrix and the stability of the fossils you are working on. Please visit a company called Paleotools (www.paleotools.com) for the best reseller and manufacturer of precision air-scribes. All air-scribes will require an air-compressor to operate. When using an air-scribe try hard not to force the tip into the matrix too much. Let the tool do the work for you. You should not need to push too hard. If the tool makes the sound of an angry bee, you are probably pushing too hard.

J. Rock hammers and chisels- The old geologist standards. Hammer and chisels work best when used to remove overburden close to, but not actually in the underlying bone bed. With patience and practice one can develop good power and precision with them, but for most novices its best to stay in the overlying zones.

K. <u>Large shovels</u>- The best shovels for paleo-fieldwork are ones with square, straight blades and short, straight shafts. These are good for removing overburden in softer sediments. They can be used to help even up quarry walls and to remove debris piles. They generally should not be used within the bone producing layer, without extreme caution. These types of shovels also make for a great walking stick when exploring.

L. <u>Pick axe</u>- Pick axes are seldom used in the field as their weight, lack of precision, and awkwardness make them impractical in most cases. At some dig sites, where the overburden is dense and highly cemented they have application. Pick axes or long crow bars may in some circumstances also be used to help flip over medium to large plaster jackets.

M. <u>Pneumatic chisels</u>- Definitely not recommended for beginners. Air-powered hammer/chisels can aid in the removal of overburden but should not be used near any bone bed, as they have little control or precision.

N. <u>Rock saws</u>- These have limited use in softer chalk, limestone or other carbonate rock types that are laid down in plates or thin to medium beds. They are predominately used to quarry out fossil fish or marine reptiles in the Niobrara Formation.

O. <u>Jack hammers</u>- Jack hammers are useful only in the removal of dense overburden.

They should not be employed any closer than 1 meter above a prospective fossil layer, unless absolutely necessary.

P. <u>Heavy Equipment</u>- Heavy equipment ranging from small skid-steers, excavators, and backhoes are often employed when the overburden above a site is too great to be removed by hand. They are a necessity when the overburden thickness is greater than two meters. If used correctly, and monitored closely, they can safely decrease the excavation time ten-fold. If used incorrectly, they are a hazard to field workers and to the fossils. Use them with caution.

Q. <u>A Towel</u>- Hey if Douglas Adams says it's essential in the "*Hitchhikers Guide to the Galaxy*", who am I to argue?

49. When in Doubt, Keep Your Chisels, Hammers and Pneumatics Holstered. This one is pretty much self explanatory. Unless you are absolutely 99% positive you are not going to hit a fossil or the solid matrix leaves you no other option, keep the heavier tools to a minimum. Pneumatic chisels especially have no business in the actual bone bed and should only be used to help clear overburden. These weapons have little control and in the wrong hands can be devastating.

50. Use #2 Scallop Bladed X-actos… Not the Annoying, Ultra Thin,

Pointy Kind: The super sharp, super thin, ultra pointy replacement blades for x-acto knives last about 5 minutes in most bone beds. They are generally a waste of money. If they have any use, it would be for helping to cut troublesome roots or removing small grains of sand. Always uses thicker, scallop-bladed, knives with heavy duty handles.

51. Do not STAB at the Earth! I'm not entirely sure why, but I find that a lot of novices tend to use the Alfred Hitchcock, "I'm a knife wielding Psycho" method of excavation. It's as if every ice pick that has ever been made is haunted with some murderous ghost bent on destruction. It is for this reason that I seldom give out ice picks to our tour groups. Ice picks, shovels, knives, etc. need to be carefully employed. We are not trying to kill the dinosaurs they are already dead. Thank you.

52. The Fossils Didn't Wait 65 Million Years For You to Come Along and #@#$@#!!?!$ Them Up in Two Minutes! This one is also pretty much self explanatory. If you still do not understand it, please review rules 1-51 or find another profession.

53. Get As Comfortable As Possible: As you begin getting down to the business of removing matrix around your fossils it is important to get yourself into a position that is as comfortable as possible. It is true, that field technicians are often known to be contortionists as they attempt to wriggle between tangles of delicate

fossils, cutting them out of the rock. Be sure to stretch, get limber and find a comfortable place. Otherwise, you will be less focused upon the surgery at hand and more concerned about the cramp in your right leg. Use foam mats for elbows, knees and shoulders. Use knee pads if you tend to have knee problems or a weight belt if you tend to have back problems. Wear loose-fitting clothes that are light in color to keep your temperature down.

54. Take What the Rock Gives You:

When excavating, try your best to not force anything. When you have to force an overlying rock in order to move it, it means that the rock isn't ready to be removed yet. Work around it until it comes free more easily. If the overlying rock is layered, split it along bedding planes rather than trying to cut through something going against the grain. Use faults and natural fractures when they show up. If the rock comes up in large chunks, take them. If it comes off in smaller chunks, don't try and make bigger chunks out of them. If the rock is soft, brush more and hit less. If there are a lot of roots, work around them until a safer path can be found. If one area of the dig is difficult or complex, unless circumstances demand it, avoid the area and work the easiest stuff first. Often, once the easiest matrix is removed it becomes easier to remove the harder stuff.

55. Too Much Torque is a Bad Thing:

Torque is when you use rotational forces to pry or twist matrix or fossils in order to remove them. If you have a large chunk of hard matrix perched over top of a softer rock with your fossil, try to pry it as little as possible as the torque will cause the backside of the hard rock to break into and possibly damage the fossil underneath. As a general rule, if a piece of the matrix does not lift straight up and away from the fossil and you have to pry it to move it, then leave it in place. If you are following rule #53, removing the easiest stuff first, more than likely, the chunk will come free as you work around it.

56. Smooth Even Pressure When Probing For Fossils: Rule #51 and #56 go hand in hand. You do not want to stab into the earth when probing for fossils. Erratic speed and pressure in blade strokes, irregardless of the tool, will inevitably cause damage to a fossil should any be nearby. You need to use constant uniform pressure with level and even strokes. If you have successfully mastered this technique and are using the correct tool, should you hit something while probing, your blade will stop and not go straight through the specimen.

57. Remove the Matrix or Leave Attached? There are two main schools of thought on this subject. 1) Some workers prefer to remove as little rock from the surface of the bone as possible, probing around the specimen for its maximum dimensions. These workers believe that the specimen is best suited to be prepared in the lab and not the field and that removing them from the ground as quickly as possible, with little interruption of surrounding matrix is best for the bone. The theory here is that it is safer and easier for a worker to prepare a specimen when in a controlled and more comfortable environment, like the laboratory. They believe that any mistakes made will be easier to correct in the lab. 2) Others prefer to remove as much of the surrounding matrix from the upper surface of the bone as possible in the field. These workers believe that the best way to preserve the bone is to prepare as much as possible while it is still in place. The theory here is that since bones

are often highly fractured, it is often necessary to stabilize them prior to transport. To do this, you need to remove the overlying matrix and glue the specimens in place. If done correctly, the specimens will suffer less damage and the quicker and safer it will be to prepare in the lab. Randomly probing for the maximum dimensions of the bone may cause accidental damage to the bone and other bones lying nearby. Also, once the rock is removed from the ground and shipped to the lab, it dries out quickly and typically cracks. This may act to further damage any bone within its depths. Without removing the matrix it is often difficult to accurately map the bones in the field as well.

Both strategies have merit and which one you decide to use depends primarily on three factors. 1) What is the type of matrix: Is the rock surrounding the bone hard and solid, such as a limestone, concretion, slate, or strongly cemented sandstone? If the rock is hard, strategy number one is best suited for the bone. If the bone is surrounded by a mudstone, siltstone, shale, or loosely consolidated sandstone, where removal and subsequent jolting can further loosen broken chunks of bone and matrix, technique number 2 is best suited. 2) What are the environmental conditions expected for the dig: Are the conditions uncomfortable, either terribly hot or cold, are insects a problem, are you expecting an abundance of rain, hail, or other damaging weather conditions. If you answered yes to any of these questions strategy number one is probably the safest bet. If, however, conditions are mild and pleasant, technique number 2 could work out the best. 3) Miscellaneous factors: These include things like time limit to excavate, preservation of bone, safety of field workers and other factors. When in doubt, if there is anything which causes one to be uncomfortable or rushed in the field, or the bone "skin" is in poor quality (IE. "punky" or weathered prior to fossilization- primary weathering) technique number 1 is best. If there is plenty of time, the bone is in good shape, and the matrix is removed easily, technique number 2 is best.

58. You Do Not Always Need To Remove Every Last Grain of Sand in the Field!

Many novice field workers, when using technique #2 from Rule #57 above, catch themselves trying to "prepare" a fossil skeleton in the field. They try to pick off every last bit of matrix, then glue all of the pieces together (and get it cleaned up as best as they can) while it is still in place. Generally this is not a good thing. You do not need to prepare the specimen; you only need to get it out in one piece. Therefore, you only need to remove the matrix from the top, not necessarily the sides. You do not have to have it cleaned, just visible. You are trying to find its maximum extent and trying to make sure it is not twisted together with other elements; not trying to make it museum ready. The excess glue and the stuck chunks of matrix should be removed in the lab. Remember, you only have so much time in the field; you need to make the most of it.

59. The More You Brush, The More You Will Find:

This rule is so important, I almost put it in the top 20. If you can't see what you are sticking your blade into, you are bound to miss things and damage others. Everything from associated bones to microfossils, to important sedimentary structures, to taphonomic and geologic observations can be found by frequently brushing your site. Too many workers under-utilize their brushes, focusing more on removing larger pieces of matrix as fast as possible and less on actually looking at those pieces. Brushing adds a significant amount of time to your excavation, but it is always worth it in the end.

60. Short, Choppy Strokes in One Direction When Brushing: Watching a
beginning field crew member work a brush has got to be
one of the funniest things us paleo-geeks get to see.
Generally, a newbie will randomly brush at the earth with
long, random strokes, chaotically pushing and whiffing in
every direction sending dust and bits of rock back and
forth over the exact same spot, over and over again, until
the dust settles... pretty much where it started. Then they
wonder why they still can't see anything clearly. When
using your brush, use short, choppy, uni-directional
strokes, brushing the dust and matrix into debris piles.
You should sweep the dust and debris toward you and
away from the quarry wall or the main bulk of the fossils
you are trying to get a better look at. Never push with the
brush. Never stab or swipe with the brush. Try hard not to
sweep left, then right, north, then south, up, then down
with the brush. Simple, short, choppy, even pressure
strokes, away from the fossils and toward the debris pile,
will always work better in the long run.

61. Fossils are like Icebergs- Never Assume you are seeing the Entire Thing: I have worked on several dinosaur skeletons
that were initially very difficult to interpret. Fossil bones
partially encased in matrix, broken by fractures and
riddled with roots can be confusing beasts. Sometimes
they dive deeper than you would expect. Sometimes they
curl in directions you didn't predict. In general, once you
know what type of element you are working on, you know
which direction it's heading. Once you know it's a rib,
you can generally tell which way it's going to curve. Once
you know it's a femur, or a flipper, you know
approximately how big it's going to be and where it is

going under the rock. Sometimes though, the bone will confuse you. You think you are working on a rib, but it turns out to be the edge of a pelvic bone. Or, you think you have a vertebra and all of a sudden it twists in a direction you didn't think possible and you realize you are working on a skull. Remember, you are not superman/superwoman. You cannot see into the rock. Be mindful, that even though you think you know what you are dealing with, you could be mistaken. If you assume too much and then rush, you could damage your specimens irreparably.

62. Gently Blow Off the Dust: Sometimes
even a brush is too much for a delicate fossil. The sweeping motion can lift off fragments and hurl them towards the debris pile where they are seldom found again. When the specimens you are working on are extremely fragile it is okay to gently blow on the fossil so you can see what you are doing. This is often done in unconsolidated sandstones.

63. Keep Your Dirt Clean and Your Site Organized: This rule probably sounds
pretty silly especially since you are digging in the dirt all day long. I mean, how do you really keep a dirty, dusty dig site "clean"? Well, a "clean" dig site is one where workers have brushed all dirt and debris away from the main specimen(s) and there is a clear distinction between what is rock, and what is fossil. Pathways around the specimen are clearly defined so that everyone knows where they can step and where they dare not. An organized site is one where everyone's set of tools is within reach and not randomly scattered amongst the bones and other co-workers. An organized site is one where all of the heavier tools and supplies are located in

one central location and not randomly scattered about. A clean, organized site is one where debris piles are moved in a logical direction away from the site instead of simply tossed in random intertwined piles. A well organized site is one where everybody can easily see where they have already been and where they are going next. One where the quarry walls are well defined so that you can clearly see the stratigraphy and can note major sedimentological changes when they are encountered. If you do these things, expect the team members to get along better, expect the loss of fossils and equipment to be minimal and the chances for error to be slimmer.

64. Set Field Tools Above You: When you are engaged in an excavation always set your field tools and glue bottles above you and within reach. Stick x-actos, dental picks and sharp knives, blade-side down into the overlying or underlying matrix. This prevents tool loss and makes it easier for the workers to grab their tools when they need them. It also prevents anyone from accidentally sitting on their own x-acto blade- OUCH!

65. Switch Tools Often: As you are excavating, be sure to keep rule #48 and Rule #65 in mind and switch your tools often. If you do nothing but work with a dental pick all day, you are probably not making the most of your day. If you do nothing but work with a trowel, it's a sure bet you are missing a lot of things along the way. You want to start with heavier tools and remove matrix above the specimen, then switch to an ice pick or X-acto as you get into the bone bed, Then switch to brush, then X-acto, then brush, then dental pick, as you get next to a fossil. Keep switching tools as often as the dig warrants. Don't get stuck with the same tool all day.

66. Always Work From the Top Down In Tiers:

Some introductory field crews have a tendency to come at a quarry wall from the side. When they do this they end up digging horizontal tunnels which they can't really see into very well. Sometimes this practice can lead to leaning quarry walls with a nice overhang, which always threatens to collapse by the end of the day. Digging horizontally can be dangerous as most fossils lie horizontal, along bedding planes rather than cross cut them. When you come at the quarry wall from this orientation you are attacking the shortest plane of the fossil (in most cases). This can often lead to accidental damage. At some locations this approach may be your only option, especially if you have a high, nearly vertical quarry wall. If you can get away with it, however, it is far better to create tiers, generally a half meter to a meter deep, and peel off the layers one at a time, brushing constantly. As you work from the top down, you attack the bones in the vertical plane rather than the horizontal which gives you a better chance of seeing long bones like ribs, limbs, pelvic and pectoral bones, or bones lying flat, like skull elements and vertebrae that are twisted downwards. This method also helps with microfossils and enables you to get a real good look at the overlying matrix for miscellaneous structures that might prove important.

67. Square Up Those Quarry Walls:

This rule is not one of the most important, but it does have some benefits. Squaring up, or shaving quarry walls, is what you typically see at most professional archaeological excavations. The overlying rock is cut into tiers (see rule #66) that are squared off from one another. This does more than make your site look organized and clean. This is a great way to map the stratigraphy of the bone bed, and can aid in mapping, particularly if you are using the grid technique.

68. It's Amazing What One Inch of Rock Can Contain: You would be amazed by the quantity of fossils that a single one inch layer of mud will conceal! This goes for invertebrates, vertebrates, plants and the like. When excavating, try to break all large debris chunks into at least fist sized pieces to ensure you have collected all there is to collect. In areas where you are dealing with a lot of microfossils you might even want to take the matrix down to quarter sized chunks, screen it or bag it for later micro-fossil analysis.

69. Glue Cracks As You Go: Fossil vertebrate and invertebrate remains have often been buried under thousands of feet of rock, for millions of years. As a result they are often highly broken and fractured. These fractures will need to be glued in place to prevent further damage and the loss of tiny fragments. There are many different kinds of glues available on the market for paleontological use. The most commonly used type of glue today, is cyanoacrolytic super glue. Each brand of super glue (Paleo-bond, Star-bond, Handi-bond etc.) has variations in viscosity (thickness) ranging from super-thin, watery-glue options, to super thick, resin-like glue options. The watery glues are best suited for areas with abundant crushing and fracturing. The thicker glues are best suited for larger breaks and heavier bones.

When you encounter a crushed or severely broken bone, be sure to place a few drops of thin glue in those fractures before removing any additional matrix. You should try to remove as much dust and debris from the fractures as possible by lightly brushing with a small paint brush or blowing gently on them.

70. Try to Keep as Much Glue off the Matrix as You Can:

Many first time diggers have the tendency to "over glue" a specimen, squirting ounce upon ounce of watery "penetrate stabilizer" super glue into each and every fracture. They squirt it on the surrounding matrix. They pour it over the unbroken surfaces. They basically go glue-crazy. This is a big no-no. Yes you want to follow rule #69 and glue the fractures as you go, but a few drops should be sufficient. Spilling glue all over a fossil bone will triple your prep-time, often cause damage to the bone skin and waste a hell of a lot of glue (do you know how expensive that stuff is?!).

71. To Glue or Not to Glue- That is The Question:

Some fractures are so severe and have so much separation that gluing them in the field would be a mistake. For these breaks, assuming the edges of the breaks are solid, you should remove the two pieces separately and wait to glue them back together again in the lab, after the fracture has been cleaned out and re-set. As a general rule, think of it this way... if the bone chunks are thick and solid, they are not splintered, sheared or powdered, if they are heavy and the separation between the pieces is large, if it is obvious how they fit together, consider NOT gluing them in the field. If the bone is highly fractured, crushed, splintered, shattered, punky, powdery or thin and you know damn well that if you do not glue it, you will loose it or confuse it, then glue it as you go (then... apply a plaster jacket, cross your fingers, and make a note in your field journal to give that prep job to someone else!).

72. To Activate or Not to Activate- This is Another Question: Anyone who has ever had a back room tour of a fossil prep-lab has seen fossils in jackets, surrounded by matrix with a bizarre greenish tinge. The color is ugly, distracting and unmistakable. This is caused by a chemical reaction between copper and iron molecules in the matrix and a chemical spray, called "Activator", used to instantaneously harden superglue. The majority of time, it is both unnecessary and unwise to use activators in the field. The glue needs time to do its job and soak into porous, punky, nasty broken bones. It is better to let it air dry, without the activator, unless time is of the essence. Generally, in the field, time is not that big of an issue, especially if you are gluing the bones in place. Too often new diggers will pour glue onto a specimen and then immediately reach for the activator. Try to keep that to a minimum. A mixture of water and bleach, I am told, is the only thing that will remove the green tinge later on.

73. Try to Keep Your Glue Cool: Glue bottles last longer when they are kept in a cool, dry environment. That's tough to do when working in high summer temperatures, but try your best. Place all large bottles in a small cooler with icepacks to keep them in optimum condition. Keep the cooler in the shade.

74. Find the Bones Maximum Extent Before Trying to Remove It: When excavating around a fossil bone make sure you have found its maximum extent before removing or jacketing it. Bones can twist and turn in all sorts of directions and they can be piled on top of one another at bizarre angles. Try

very hard to keep as much of the original orientation together in one block if possible. (See Also Rule #61.)

75. Never Undercut a Bone: When working around skeletal remains, be aware that some bones will need to be jacketed. Most will not be in good enough condition to remove separately without the aid of plaster and burlap. Do not undercut a bone unless you are absolutely sure that you have found its maximum extent and that you can pop it out cleanly. Do not undercut around a fossil unless you are sure you have enough time that day to begin laying out the plaster jacket. If you undercut something, then run out of time, expect to find a collapsed chunk the following morning.

76. Never Pull a Root: Fossil bone is often very porous. Ground water may be trapped inside those pores, and as the specimen weathers closer and closer to the top soil, roots seek out that moisture and grow right through your delicate treasure. As the roots continue to grow, they sometimes trigger additional fracturing and splintering. They begin to dislodge pieces. They work their way into the fossil "marrow" causing it to become powdery. They cause a chemical discoloration to once pristine surfaces of bone and teeth. Roots are one of the most annoying things encountered on a dig site and it sometimes takes days to work around them. Naturally, many workers are inclined to begin pulling on the roots in order to remove them, but this can be a big mistake. As you pull on the root, the matrix and the other hundreds of root tendrils that are connected to it come as well. This can severely damage the bone that it surrounds and runs through. Roots need to be cut (using scissors or knives) or burned away, rather than pulled. It takes a lot longer, but you and the fossil will be better off.

77. Excavation for Fossils Involves More Than Just Sight; Hear the Fossil, Be the Fossil:

Okay this does sound pretty metaphysical doesn't it? You are probably picturing Chevy Chase in the movie "Caddy Shack", out their doing some sort of far eastern meditation, hovering over the fossils and praying to Buddha. That's not quite what I mean when I say "Be the fossil". What I am attempting to explain is that too often, introductory students use only their eyes when excavating. There is clearly a visual component to excavation, but there are other senses that you need to employ as well, or you will miss things. If you are working with nice uniform even strokes (Rule #56), you will also be able to "feel" your way through a bone bed. The x-acto knife, trowel or pick, will become an extension of your fingertips. When you hit something solid, you will feel it. If you are careful, you can back out and go around the spot until you carefully find out what it is. Sometimes as you are working, you will "hear" the fossil, as you accidentally scrape past it. If you are going too fast you might "hear" a crunch, or a grinding sound. Workers who do nothing but talk at a dig site or listen to loud music, can crunch through dozens of small, soft bones without even noticing it. By the time their eyes "see" the broken fragments in their debris piles it is often too late. So, the moral is: its okay to use some of your other senses when excavating… it doesn't make you a new-age guru.

78. Heavy Equipment? Be Extremely Careful:

As stated in a previous rule, the use of heavy equipment at a fossil dig site is essential to the speedy and complete recovery of large vertebrate remains. In the right hands, a small skid steer is a work of art, pivoting on a dime, scraping a centimeter of overburden at a time. In the wrong hands, it is like a rampaging mechanical T. rex, thrashing about in a china shop, being controlled by a four year old boy (i.e. DANGEROUS)! Dangerous to both the fossils and your crew if you are working near the operator. Make sure that your equipment operator understands what you expect them to do. Clearly denote areas where they can go and where they can't. Clearly mark the maximum depth you want them to dig. And always…. always, have someone watching the slices for traces of fossils in the overburden. It is highly possible that in removing layers of rock above a known bone bed, that the operator will accidentally hit another one above it. Be ready for that possibility and be careful.

79. Use Aluminum Foil as a Separator:
After you have pedestalled your fossils and prepared them for a plaster jacket, you will need to place some form of separator between the exposed bone and the coming plaster of paris. In the past, field workers used a variety of things as a separator, including wet newspapers, toilet paper, straw, leaves, rags, small interns, etc. Basically, whatever they could find. Whereas toilet paper is still sometimes used in limited or emergency cases, most of the other forms have gone the way of the Dodo. The best separator that prevents the plaster from leaking onto the fossils (and possibly damaging them or making them impossible to remove cleanly from the jacket), is good old aluminum foil. Be sure to use the extra heavy kind… not the cheap single layer kind.

80. Before Jacketing Go Deep: Remember
that fossils are sometimes like icebergs? It is for this reason that you need to carefully pedestal and undercut a specimen block, going as deep as necessary to determine

conclusively that there aren't any surprises underneath. Twists and tangles of bones obscured underneath could easily make flipping a shallow jacket a sad day on any dig site.

81. Warm Water Dries Slower and More Brittle than Cold: When mixing your plaster of Paris use cool or cold water and mix to the consistency of runny mud for the best results. If your water is too warm or hot from sitting outside in the sun all day, the plaster will dry slower and be more brittle. If your mixture is too thick, it will set up faster than you can lay it. If it is too watery, it will not have the strength needed to protect the bones.

82. When Jacketing, Your First Layers Should be Wrapped Around the Bottom Edge, Not Over the Top: To ensure the jacket has a tight grip around the bottom of the specimen, your first layer of plaster-soaked burlap should always be wrapped around the bottom edge first, when possible. Subsequent layers should then criss-cross over the top from various directions. Depending on the size of the jacket, you should then do another layer around the bottom edge and a final larger piece over the top.

83. One Layer of Plaster and Burlap is Usually Not Enough: I'm not sure why, but too many introductory students and even some seasoned vets have a tendency to go really light on their jackets. One layer of burlap is seldom enough to protect a

fossil bone. One layer without any cross cutting sheets will have a tendency to break over time. If your jacket is going to weight more than five pounds, you had better use a minimum of two layers criss-crossing one another.

84. 10-12 –Inch Strips of Burlap Cut on an Angle Work Best: Every field crew

has their own unique methods for cutting burlap and making jackets and each thinks their method is the best. For me, call it OCD (obsessive compulsive) if you like, but I can't stand ugly jackets with frayed ends, uneven plaster and sharp edges that tear your forearms apart when you carry them. It is for this reason that I would argue your burlap needs to be cut uniformly, approximately 10-12 inches wide and cut on an angle to prevent fraying. The plaster should be of a uniform consistency, smooth, without lumps, and as dirt free as physically possible. As you place the plaster soaked burlap around the pedestal be sure to fold the bottom edge up (creating a double layer around the bottom edge. This minimizes sharp, stringy edges and makes one hell of a pretty, and more importantly, safe jacket.

85. Label Those Jackets: Once the plaster has

hardened and the jacket has been flipped over, take a ¼ inch wide paint brush and some brightly colored acrylic paint and write on the cleanest side, the "site name" and the field numbers of the important bones contained therein. Preparators and supervisors will want to know what each jacket contains before they open them up and begin preparation sometimes months or years in the future.

86. Use Wooden Braces to Stabilize a

Jacket: For heavier jackets, it is usually a good idea to brace the sides and bottom with wooden boards or metal framing. This keeps the jacket from breaking in the event you used too few layers to hold the jacket's weight. In a best case scenario, this will create a level base that the jacket can sit on, which prevents it from rocking during transport or preparation.

87. Screen Your Debris Pile: Everyone

misses the occasional microfossil or accidentally brushes away the occasional bone chip. Remember you are not god, nor superman, and you will break stuff. It is for these reasons that your debris piles should be sifted to ensure that you have recovered everything humanly possible.

Some of the most amazing things I have found over the last 12 years were less than the size of a penny- some the size of a pea. Fish bones, pterosaur teeth, lizard jaws, mammal bones, frog and salamander skull elements and even rare dino bones and teeth can all be found by screening. The big boys were not the only things alive back in the day. Screening will ensure that the smallest of vertebrates are recovered safely. These little guys can often tell us far more about the ancient world than the bigger specimens. They help to fill out our ecological picture, tell us a bit about the environment and the climate and other important features of that ancient landscape.

88. Cover an Active Dig with a Tarp at the End of Each Day/Carefully Dig Drainage Trenches: If you are

working on a large vertebrate fossil, be sure to cover all of the exposed elements with a tarp at the end of each day. In the badlands, the weather is often unpredictable. Sudden rain storms or hail storms could easily damage exposed elements. Rainwater from those storms can easily drown a site making it unworkable for days. After all, no one wants to be taking a slimy mud bath while they are trying to carefully excavate a delicate, 10 million year old *Smilodon* skull. If the tarps are large enough and drainage trenches are built around the site, you can sometimes avert playing in the mud. We neglected to do this once on an Edmontosaurus dig site and created a nice little pond some two feet in depth that lasted the better part of a month.

89. Fence Off a Dig Site If Necessary:

Cows are inquisitive, brilliant animals… that's why we love to eat them! If you are working in an area with cattle, expect your bovine neighbors to periodically investigate your dig sites from time to time. Deer, antelope and occasionally other wild game might also find your new locality a point of great interest. For these reasons, you might want to consider placing a tight, barbed wire fence around important sites, especially if you intend to leave certain fossils in the ground for extended periods of time. You certainly do not want a herd of curious cows to trample the specimens you have been working on for weeks. In other remote areas, where human visitation sometimes occurs, it may also be a good idea to clearly mark the site with a fence and post signs which clearly outline the penalties for vandalizing and destroying the specimens contained therein. It won't necessarily stop wanton destruction, but it will sink the culprits should they be caught and sent to trial. They won't be able to pretend ignorance.

90. Remove All Bones Prior To Winterizing A Site If Possible: Some

dig sites require several seasons of work before they are officially closed up and your crews move on. If you reach the end of field season and you know you will not be able to remove all of the specimens before the first snows begin to fly, then you will need to cover them with a tarp and re-bury the site. The tarp and the debris overlying the fossils will give them limited protection from the elements and keep at least some of the winter's precipitation from freezing-thawing-freezing-thawing your treasures to dust. When opening up the site the following season, be careful when removing the tarp as lots of little creepy-crawlies may have found a new home over the winter under your tarp. Rattlesnakes, spiders and scorpions often do this.

D. Documentation:

91. Put One Person in Charge of All Mapping and Logging of Fossils:

Rule number #5 tells us that it is of utmost importance to document everything you can about a fossil locality. Rule #91 tells you that the majority of that work should fall to just one person. One person doing all of your mapping and logging ensures that there is uniformity to the data collection. Measurements are taken in the exact same way for each specimen. They are each logged in the same fashion. Lines are drawn in the exact same way for each grid. Subjective terms are uniform and comparative and have greater reliability. There might be errors, but at least with one person making them you usually have uniformity of error. These are far easier to catch and fix. If you have ten different people all mapping bones or taking measurements, or describing rock units or structures, you have ten different sets of sometimes widely differing variables. I'm not a statistician, but this clearly increases your margin for error by a large factor.

92. It's Not a Bad Idea to Have A Backup- Just in Case:
Whereas, rule #91 suggests that one person be in charge of all mapping and logging of fossils, rule #92 tells us that it couldn't hurt to have at least one back up. This should be a person who knows exactly how the first person took the measurements or logged the fossils. They should know what the principle mapper meant when he or she wrote that the bones were in "good" quality, or were "reddish brown" in color. You would hope the principle mapper is not color blind and has some quantitative reasoning for labeling a fossil "good" or "poor" quality, but just in case subjectivity enters too much into the approach, you have a second set of eyes that could interpret the notes (just in case!). Your principle documentarian may get sick, injured or decide to resign; without a backup you would be in trouble.

93. Keep a Daily Journal:
As stated in Rule #5, keeping a journal is essential to good record keeping. A waterproof or hard back journal is best and can be purchased at any bookstore, though any old notebook will do in a pinch. Your field journal is important, because it is your record of the actual dig. People are forgetful creatures. I know I am! It is easy to forget things when you are rushing about trying to excavate a specimen. Writing the details forces you to slow down, makes you think and keeps a record of the site's activities that can be useful in the future analysis of the site. If you do not write it down you will lose information. It will probably make you look pretty silly when your peers start drilling you with questions about your new find and you can't recall the answers and there is no record of it anywhere.

In your field journal you should record all aspects of a dig from start to finish on a daily basis either during the excavation or at the end of each day. You will record things like the date, time, weather conditions, the name of the site (usually a two to three letter abbreviation; EX: "John's Mosasaur" = JM site; EX: "Dalton Ranch Brontothere"- DRB site, "North Caleb Buttes Bone Bed"= NCB site), location of the site including GPS coordinates, names of each field worker, descriptions, orientations and measurements of key fossils, petrology (rock descriptions), rock formation and age, stratigraphy (how the rock layers are laid down), the sites physical description, the history (provenance) of the discovery, the research goals or questions and other important observations encountered at the dig. When writing think about the following questions; Who(?), What(?), Where(?), When(?), Why(?) and How(?), then answer each in as many details as it takes. Use as many adjectives as necessary to answer those questions. For example, was the rock a sandstone or a limestone? What was its color? Were their any distinguishing structures in it? What was the grain size? Were those grains angular or rounded? What type of rock was above it and below it? Who conducted the petrographic analysis? Was a sample collected? What was it labeled? Where was it stored? When was it removed? Who removed it? Why was it collected? How was it collected? If you have not used your journal to answer each of those questions, for each important aspect of the dig your journal may not be considered complete. One of those questions might come back to haunt you down the line. For example, one of your preparators frantically informs you that a large piece of mammoth femur is missing from its plaster jacket. They ask you if you remember what happened to the piece and you say... "hmmm... that was five years ago, but oh... wait, I have my journal". You ask when the femur was collected and find that date in your journal. You read the entry for that date and scan to the portion where you have described the removal of the femur and the name of who removed it. You quickly discover that the missing fragment was fractured, and removed separately in order to lighten the jacket. You discover that it was labeled "such and such" and was placed in a certain box labeled "such and such". With your new information acquired from your journal, you quickly take your frantic preparator over to the box buried in storage, containing the missing piece. Problem solved. Imagine the same scenario if you hadn't written that in your journal!

Make drawings or sketches in the journal if it is easier to describe something visually. Make notes and include a scale on your sketch. Use color if it helps.

94. Keep a Fossil Inventory List: A fossil
inventory list is basically a logbook used to keep track of your collected specimens. It contains key information about each one of the fossils or elements you have collected. It provides an identification number for each, which should be written in the logbook and written on it's corresponding fossil. The number should have the abbreviation for the site name and a sequential number for the individual element. The logbook also tells what each element/fossil is and from what genus or species if known. It lists each specimen's map coordinates, orientation and vertical or stratigraphic position. It lists who found the fossil and on what date. It lists when the specimen was removed from the field and where its current location is. It will sometimes list descriptors such as color, quality, dimensions, preservation, condition, or notes describing important pathological or taphonomic data. It is your complete record of everything collected on-site. If it is done well you will have minimum misplacement of specimens or data. It will aid preparators in planning. It will aid supervisors and scientists in analysis of the site.

95. Provide a Proper Site Description:
The site description is one of the most important things you want to include in any paleontological field journal. It includes things like the exact location of the dig site, the

nature of the terrain, the distance and direction to known, semi-permanent landmarks, the nature, size and orientation of the outcrop, the degree of vegetation, the preliminary description of the stratigraphy, the strike and dip of the beds, the nature and scale of the debris field, the thickness of the bone bearing horizon, the initial condition of the weathered remains and other factors. Much of this data will be useful when analyzing many important research aspects of the dig. Another purpose of recording this information is so that future workers will be able to come back to the EXACT location should further work need to be done at a much later date. We would all love to know the exact locations of some of paleontology's most significant finds. As stated in a previous rule, these sites often need to be re-evaluated and re-excavated sometimes dozens of years after the initial discovery. Keeping a record of the site ensures that future work can be done. If you would like more information on writing a site description please see my paper "Field Journal Documentation Procedures Part One: The Site Description" (Stein, 2007), on the *Journal of Paleontological Sciences* website at (www.aaps-journal.org).

96. Provide A Site Sketch: It often helps to draw a quick sketch of the outcrop in your journal. The sketch can quickly show where the specimen was eroding out in regards to the outcrop, pinpoint nearby landmarks, and show things that a photograph might not. Be sure to include an approximate scale and exact date in your drawing.

97. Sedimentary Layers are Like Pages in A History Book- Learn

How to Read the Pages: Each

sedimentary layer is like a page in a history book. The pages are just crushed, folded, turned sideways and composed of solid rock! By analyzing each page or each layer and learning how to read the clues contained within them, you will be able to interpret the entire geologic history of an area.

The geologic Law of Superposition tells us that the layers on the bottom are, under normal circumstances, older than the layers on the top. So, instead of reading from left to right, as in a traditional history book, geologists must read from the bottom to the top. On your site sketch, you will want to draw each one of the pages (beds) or similar groupings of pages (units), (depending on the scale and precision you are after), and note why each is different from the others. You will note things such as thickness of the beds, lateral consistency of the beds, shape of the beds, important sedimentological structures (like ripple marks, mud-cracks, worm burrows or cross beds) found in the beds, color, rock types, fossils and any other features of the beds that will help you to read the pages.

Each one of these beds or groupings of beds will represent an ancient paleo-environment, such as an old lake or stream, a floodplain, an ancient river system, an estuary, a swamp, a beach or even a shallow sea. Based upon the clues and characters of each layer (shape, composition, chemistry, color, texture, structure, etc.), you will need to determine which bed equals what paleoenvironment.

Like many old, old books, occasionally we find areas where a page has been torn out. These "missing pages", also known as unconformities, represent a period of time where erosion was greater than deposition. There is a never-ending struggle between erosion and deposition. In areas where we have deposition, we eventually find sedimentary rock layers. It areas where we have erosion, we lose rock layers. When we encounter a bedding plane that is irregular in regards to an overlying one, we have an erosional surface or unconformity, and thus, a missing page.

The sequence of the paleoenvironments and the unconformities can help us to determine the region's geographic, tectonic, biologic and climatic history. Collectively, this stratigraphic section, this history book, can tell us many things about our main fossil specimen and the main fossil specimen can tell us many things about the area's history. Read together, they tell us the complete story.

98. The Rock's Story- Do Your Petrography: Your main fossil specimen is not the only thing you should collect at your dig site. You should also be sure to collect several rock samples from each one of the beds above, below, and within your dig site. Save some of these samples in air-tight zip-lock baggies for pollen or other analysis. Use some of the samples to do detailed petrographic descriptions of the rock types in each bed. When it is time to do your detailed petrography, carefully study under magnification the rock samples to determine their precise composition, color, grain size, angularity of grains, type of cement, and exact rock classification in order to help determine their paleoenvironments of deposition.

99. Draw A Stratigraphic Cross Section: Using your site sketch and your petrography, the next thing you should do is to place your specimen's position in a detailed stratigraphic column. The column is typically larger in scale and tries to show where your specimen is in relation to the entire rock

formation and to the other rock formations above and below the unit. This will help you place the age of the specimen and the dig site. This will probably take some time as you carefully measure how far below the overlying contact you are and how far above the underlying contact. There are many professional surveying tools that can help you to do this, or you can simply use a Jacobs's staff, a tape measure, and a very good eye. Once the section is measured and the geology is noted along the way, draw a stylized column pointing out major changes in lithology, important structures, index fossils and of course your dig site.

100. Mapping- Which Technique to Use? Mapping your fossil locality is also very important, particularly for vertebrate specimens, as the positions of the bones and the strata can reveal an awful lot about how the animal died and why it ended up where it did. This sub-discipline of paleontology is called taphonomy. Maps can be drawn in various ways using various techniques. These include grid mapping, radial-arc mapping, photo negative mapping, mapping using sophisticated GPS devices and lasers, and several others. By far the two most common types of mapping for vertebrate dig sites are grid and radial-arc.

Grid mapping involves surveying a physical grid over your dig site, usually in meter square lines. This is done by first surveying in a straight baseline using a compass or other survey tool and placing stakes in the ground every meter along that baseline. Next, you shoot additional lines perpendicular to your baseline, also placing stakes or survey markers every one meter. When you are done, you should have markers representing the corner points of your grid. String can be tied to each of the corner points to form the grid boundaries or a moveable 1 meter square frame can be built and moved along the lines. Specimens are then hand drawn on a mapping board, based upon the specimens' relation to the grid lines. Though this method is very accurate and the one most commonly used in both archeological and paleontological sites, there are a few problems with it you should be aware of. First, it's very difficult and awkward in many cases trying to work around a tangle of grid lines. Second, the quality of the map depends highly on the skill of the map maker and how well they can eyeball (or even measure) the bones in with the grid lines. The process is also very time-consuming, and in most cases, you want to jacket and remove the bones as soon as they are exposed rather then having to wait for a survey and a detailed sketch. The last problem is that all maps must be hand drawn in the field, where they inevitably get dirty, torn and mangled, and then redrawn at a later date.

The second most common technique is the radial-arc technique. This is achieved by surveying in a straight base-line, usually on a north-south, east-west trajectory and placing three or four fixed stakes at 5 meter intervals. These stakes remain for the entire dig and are sometimes even left at the site for future reference. Workers then measure the distance between one of the fixed points to a particular point on the fossil you are trying to map. Usually it is measured to the center point or the end point of a fossil. This distance is recorded in your log book. A second measurement is also taken from another fixed point to the exact same spot on your fossil. That is recorded. The orientation of the long axis of the bone is recorded with a compass and the bone can be removed. The map is drawn at a later date, based upon the measurements taken for each fossil and the photographs of the specimen in-situ. The map-maker draws a straight line on the map representing the baseline. Each of the fixed points are drawn based upon the chosen scale. Using a drafting compass, the map maker scales down the measurements recorded in the field and strikes an arc with the compass. They then take the second measurement from the same fixed point, scale it down, and strike a second arc. Where the two arcs meet is the location of the bone. The Radial-Arc technique is quick and easy and produces fairly accurate maps, but relies on the careful measurement of coordinates in the field. If the measurements are not correct, the position is going to be way off and there is no going back.

Whichever methods you employ, be sure to document the precise locations of your fossils in both the horizontal and the vertical dimension. Note the method you used and store it for future reference.

101. Map Everything: When you map your dig site, you will want to look, not at just the vertebrate remains, but also any other strange features, rock structures or microfossils you encounter. If you notice a fault running through the dig site- map it. If you see a sudden lithological change around the skeleton, map it. If you see gastroliths or microfossils near the belly- map them. If there are large carbonized plants, palm fronds or logs, map them. Map everything.

102. Measure Twice Record Once: When you are mapping it is always a good idea to double check each measurement to make sure it is accurate. If in doubt ask another staff member for a second opinion.

103. Measure the Vertical As Well as The Horizontal: When mapping, be sure to also have some form of vertical control. If it's high in the bone bed or low in the bone bed mark that in your log book and journal. If it is above or below the main bone bed, by how much? If you have already worked out the stratigraphy, place each element within that stratigraphy. Watch out for steeply dipping fossils, fossils that overlie one another and

bone imbrications which if not noted, can get confused on your map.

104. Photograph Everything: Just as you should map everything, you should also photograph everything. Take pictures of the dig site from various angles making sure you note the direction in the photograph. Take pictures of the people working at the site. Take pictures of them jacketing specimens and mapping elements. Take detailed, close up shots of each specimen before it is removed. Take close up pictures of any structures or oddities you observe. Photograph everything! Again, once the specimen is out of the ground, there is no going back.

105. Use a Scale in All of Your Photographs: A picture is not worth a thousand words unless you also include a point of reference. USGS photographic scales are simple, plastic or metal, metric and standard scales that are both inexpensive and widely available. If you do not have one of those handy, a quarter (exactly one inch across), a pen knife or a brush can also be used. For larger images consider using another person or a large meter staff. In a photo… size does matter. Make sure whatever scale you use, tells the viewer the approximate size of the object in question.

106. Videotape Important Dig Sites: Sometimes photographs are just not enough to truly capture the flavor of a dig site. To record certain things at a dig site sometimes it's a good idea to videotape it. Video of workers excavating at the site could be used for educational purposes or to provide a historical record of the activities that took place. Video tape of a site's fossils

could help to re-construct the orientation and position of the bones. Time lapse video is a great way to show the progress at a site. If the budgets are available, video is a wonderful addition.

107. Always Collect the Small Stuff:

We have alluded several times to the importance of collecting microfossils in many of the previous rules. Microfossils help place the main specimen in its ecological context. They help one to define the climate and ancient weather conditions. They help to show the diversity of organisms living at the same time as the main specimen. In some cases, they might represent stomach contents, predators or scavengers of the carcass. Always be on the look-out for associated fossils because in many cases they can be more important than the main specimen. As stated before, collect as much as you can by making sure all debris is screened and sorted.

108. Bag it and Tag it: Most of the time a plaster

jacket is required when removing large fossil elements from a dig. Sometimes, however, loose, smaller and better preserved things are fine without it. Any fossils that are removed without the aid of a jacket should be carefully wrapped in aluminum foil packets or zip-lock baggies. Each packet should have the abbreviation of the site name and its corresponding log number. You may even want to put the date of collection and the name of the collector on the package. Each associated microfossil, plant specimen, and rock sample should also be bagged in small zip-lock baggies or foil packets and labeled indicating their relative location to the specimen. Most of the time these fossils will be too small for mapping, but when unusual clusters of them occur pay more attention to detail.

109. Label all Boxes, Jackets, and Collected Specimens:
If you do not label everything you collect, expect at least some of the things to get misplaced or forgotten. Murphy's Law predicts that if you fail to label something, it will probably wind up being significant. Boxes of collected specimens should include a date of removal from the field. As stated in rule # 85 all plaster jackets should be clearly marked.

110. Collect Un-Tainted Geologic Samples:
As stated in previous rules make sure you collect rock samples from each of the main rock layers above and below the main bone bed. Collect several from the bone bed itself. Some of these may need to be chemically analyzed or sent to the lab for pollen analysis. To avoid contamination make sure your sample is fresh rock that has not been exposed to the air, elements or roots. Bag it and seal it quickly using as little contact with tools and greasy hands as possible.

111. Never Assume Your Initial Identification is 100% Accurate:
Until your specimen is in the lab where it can be properly cleaned up and studied, its usually best not to go blabbing to the press and your peers that you have a one of a kind species that no one has ever seen before. Thanks to the crushed condition of most vertebrate finds many times your initial identification down to the species level and even sometimes the generic level is incorrect. Be patient- save your speculations and cork your champagne bottles until you've really had time to look at, research and compare it.

112. Clean, Organize and Sharpen All Field Tools at the End of Every Season:

Oh, I am so guilty of breaking this rule! Usually, when field season is over I am so eager to start preparing our discoveries that I rush right to the prep lab and get to work. Most of the time I'm so excited, I forget to turn off the truck, let alone clean and sharpen the field tools. As a result, each and every following spring, I end up wondering what happened to all my field gear. 50% of them are missing and the rest looks like they've been salvaged from a garbage dump. Yeah, breaking this rule isn't the end of the world, but field gear costs money. If you want to save a little green for next season, follow rule #112.

E. SAFETY IN THE FIELD:

113. Boy Scouts and Paleontologists Should be Prepared for Everything:

That's right, paleontologists are boy scouts at heart. Like boy-scouts we should be prepared for every emergency the doom and gloomers can think of. You should have a well stocked first aid kit and know how to use it. You should have that snake-bite kit and tourniquet. You should have a handy-dandy Red-Cross handbook outlining what to do in case of a serious medical emergency. You should have a day or two worth of non-perishable food tucked somewhere under the spare tire, just in case you get stuck overnight. Of course, you should also make sure you have the spare tire. You should have a blanket, a battery powered lantern, some matches and some candles. You should be ready- just in case. Most vertebrate dig sites are located in remote corners of the planet. The more remote the location, the more you need to plan and the more you need to have with you in case of an emergency. Vehicles break down. Unpredictable storms wipe out roads and bridges. Socialist guerillas do rise up and steal passports (okay that WOULD be a little remote!). The point is, try to think of all the things that might go wrong and have an answer for them. Once you know you are prepared, you will have little to fear and little to worry about, if the worst does come along.

114. Be Safe. The Nearest Hospital is X Miles Away: Most field localities where fossils are found in abundance are far away from any traditional medical facilities. Keep that in the back of your mind, when the dummy in you starts thinking about climbing out on a cliff face without ropes or a partner! Picture the time you will sit in agony waiting for a helicopter to arrive, BEFORE you decide to play with the rattlesnake that just crawled into your camp. Remember Rule 114, when you tell yourself you CAN explore six more miles, in 110 degree temperatures, instead of heading back to camp, even though your water bottle dried up over an hour ago. Think about how far that hospital is before you go and try anything risky or foolish in the middle of nowhere. Be safe.

115. It's Damn Hot Out There: Sunburn, sun poisoning, heat stroke and fatigue are all constant threats while doing paleontological fieldwork. If you feel yourself getting nauseous, dizzy or light-headed, find shade, drink water and get to an air-conditioned truck as carefully and as quickly as you can.

116. Drink Plenty of Water: The average person needs to consume about 3 liters of water per day.

The average exploration paleontologist needs 2-3 times that, especially when the thermometer climbs above 100 degrees and you have miles of walking to do.

117. Use Plenty of Sun Block: There's a big yellow star in the sky called the sun. Some days it's our friend, other days it can kill you. Wear a wide-brimmed hat and loose fitting long sleeved shirt while working outdoors in the summer time and make sure you have applied plenty of sunscreen.

118. Cell Phone? Can you hear me now? No? Damn! Most of the time you will be working in remote areas like Utah, Montana, Argentina, Africa, or Mongolia. Cell coverage generally doesn't work too well in places like that. Like many paleontologists I am more adept at lighting signal fires than learning how to operate a 21st century cell phone. For many years, I resisted the technology, thinking that I wouldn't be able to get coverage in any of the areas I traditionally worked. Until one day, my lovely wife talked me into signing up for a cellular plan and taking one with me to the field. Surprise-it worked! Not everywhere, but in certain places it worked. So, even if you don't think it will work where you work, get one anyway and give it a shot. If it doesn't, try a satellite phone of some sort. You never know when you may need to call for help.

119. Always Top off The Gas Tank:
Not only are you usually far from a hospital, but often you are far from a gas station too. It's generally a good idea to top off the gas tank every time you are near any civilization with a gas station. When traveling way off the grid, be sure to carry at least one small, 5 gallon gas can

with you. That should get you at least 50-75 miles closer to home if the tank runs out.

120. Take Time to Smell the Roses:

Hey, you get to work in the best office imaginable! Don't get so caught up in your work that you forget that. Look around and enjoy the day from time to time. A relaxed field worker is a careful, safe and patient field worker.

121. Watch Out For Creepy Crawlies:

Most remote dig sites have their own host of little nasty creatures that love to bite and sting. Spiders, scorpions, ticks, fire ants, gnats, sand fleas, mosquitoes, snakes and rabid Tasmanian wombats are frequently encountered in the field. Well, okay, the wombats are not often seen these days, but, the point is, you are working in the wild. Watch out for the things that might cause you discomfort or worse. Do not stick your hand inside deep, dark crevasses that might be home to something that doesn't like you. Be careful around boulders. Walk single file with a walking stick in snake country. Make lots of noise and vibrations to let the snakes know you are present. Check for ticks after walking through tall grass. Don't play with the Gila monsters. Check your boots before you put them on in the morning. Carry spray insect repellent with you and don't, I repeat DON"T set your sleeping bag on a fire ant hill. OUCH!

122. You Can't Dig if You're Dead! I

think the cartoon adjacent pretty much says it all. Your number one priority in any paleontological quarry or exploration is safety. Try not to become a candidate for a Darwin Award.

#122- NATURAL SELECTION AT
ITS FINEST!

IV. Rules For the Laboratory

123. Preparators are Like Surgeons... Thankfully the Patient is Already Dead: Surgeons use a steady hand, a sharp eye and a keen knowledge of human anatomy to repair damaged organs, soft tissue and other squishy bits. They must have extensive training, for the life of their patients depends upon it. Preparators must also have a steady hand, a sharp eye and an intimate knowledge of various organisms' anatomies. They have to be just as careful and dedicated as a surgeon, but they don't have to deal with all the blood and gore. Nurse! Suction! Ekkk!

Preparation requires an individual to be exceptionally passionate and dedicated to their art. Often, a preparator works long, grueling hours, for little pay and little thanks in some dark university basement or glass enclosed hamster cage. They have to love what they do. If you mess up badly as a surgeon you could kill your patient and have a nice lawsuit on your hands. If you mess up as a preparator you can permanently scar or destroy a specimen that is irreplaceable. You probably will not be sued and no one will die as a result of your negligence, but you could be fired. You need to take the same serious approach as a surgeon when learning how to prepare fossils.

124. Preparators Should Also Get Field Experience: Many preparators at local museums have unfortunately had very little experience

#123- Unfortunately, early preparators left little for us to examine.

with excavation in the field. Supervisors need to try and get their team of preparators lots of field experience to help them understand some of the tricks and skills necessary to do good fossil preparation. Preparators need to understand WHY the specimens that come before them in the lab were packaged and handled, as they were, by the field staff. They need the experience of probing blindly in rock, not sure where the next fossil may show up, in order to teach them patience, focus and respect. It will help them to understand anatomy better and show them how to recognize one bone from another when they are partially or completely encased in rock. Some field experience is a must for preparators. If anything, at least it gets them out into the sun every once in awhile.

125. Fossil Preparation Takes an Artist's Touch: If you plan on spending a good deal of your time doing fossil prep, it is a very good idea to be trained in various forms of art. This includes painting, sketching, sculpting and photography. You need to be very focused and patient in order to be a good preparator. Art training will help you to develop that. Painting, gives you a delicate touch and an eye for detail. Sculpting classes will help you immensely when doing restoration, reconstruction or molding and casting. Photography will aid you when taking pictures of the prepared specimens and will even teach you a few things about composition and aesthetics. All of the above will aid you in your understanding of anatomy. Of course, that art training should be in the classical or realistic forms. If you are trained to mimic Salvador Dali or impressionists, your fossils may wind up looking like true Hollywood monsters.

126. Study Up On Your Anatomy: The

preparator needs to have a well-rounded knowledge of geology, chemistry and biology. Mostly biology. When you are removing matrix from a fossil it's a good idea to know what you are working on and where it might be heading. Too many museums rely on poorly trained volunteers who have no idea what it is they are working on. A recent surprise visit to a "professional prep lab" at a major east coast museum proved this. I asked the preparator, a very nice young kid, what it was he was working on and he could not answer me. He didn't have a clue. They simply gave him an air-scribe and some tools, stuck a random bone in his face and said "go to it". That is a terrible position to be put in. Upon closer inspection, I noticed that the nice young man was working on what appeared to be the fused atlas/axis or syncervical vertebrae, from a very young, juvenile Triceratops. That's a very important and difficult bone to be working on even for an experienced vet! Know what you are working on and understand why it is significant.

127. Be Confident: Indecisiveness and fear will

damage specimens. Cockiness and carelessness will damage specimens. There is a fine line between being confident and being over-confident. If and when you find the balance, it means that you are comfortable with the tools and equipment you have available to you. It means that you have a good understanding of what to do and when. It means that you know what you are looking at and why it is important. If you are confident, you will be able to handle most specimens that are set before you, and even if you do make a mistake, you will know how to fix it without panicking. If you are NOT confident, you may make a very poor decision that could make things potentially much, much worse. If you are fearful and scared of making a mistake- you most certainly will. You will either become so paralyzed that you won't be able to get anything accomplished or you will panic and force something that shouldn't be done. If you are over-confident, you will tend to rush and have poor judgments leading to possible damage. Know your limitations, but be confident.

128. The Right Laboratory Tool for the Job:

There are many different laboratory tools used to prepare specimens. Each lab has its own preferences, standard procedures and techniques. Many of the same tools used in the field are also used in the lab; just the setting has been changed to a more comfortable, climate-controlled environment. Please see Rule #48 for information on things like dental picks, X-actos and air scribes and their use. Below is a list of a few other tools you will have in the lab and some of their uses.

   A. Fine needles and pin vises- A pin vise is a small hand-held tool with a needle point or "stylus". The needle shapes and sizes can vary widely and are often interchangeable. Pin vises are frequently used hand tools, when working on extremely delicate, tiny or important fossils. They can be used to help separate tightly packed fossils or chisel underneath fragile ledges and overhangs of bone. They can also be used when cleaning out fractures or in areas where more invasive tools would cause damage. They are easy to use and cause little accidental damage even in the shakiest hands. They are not very good for large areas of matrix and take a tremendously long time to see tangible results. If you need to remove a single grain of sand at a time, this is your tool.

B. <u>Small or fine dental picks</u>- Lightly built dental picks are commonly employed in fossil preparation. They have many of the same uses and drawbacks as the pin vise.

C. <u>Toothbrushes</u>- Why bang when you can just brush? Sometimes all it takes to remove fine, powdery, poorly cemented sand is a common toothbrush or other lightly built brush. These are great for working on the articular surfaces of bones which tend to be very soft and easily damaged. A simple brushing is usually enough to get those ends cleaned. Hell, if it works on your teeth, why shouldn't it work on a raptor's tooth?

D. <u>Clay working tools</u>- Clay working tools like long handled, fine spatulas, picks, and the like have limited applications in fossil prep. Usually, their tips are not fine enough to do the job right, but in some circumstances they are better than the other tools. You will of course be using clay-working tools during the molding and casting phase.

E. <u>Rotary grinders</u>- Rotary grinders made by companies like Dremel or Black and Decker can be found at most hardware stores and work well in many situations. They can be either electrically powered or air powered and most labs have a mixture of both. They come in a variety of sizes and strengths and usually have about 1 million interchangeable tip options. The tips can be anything from drum sanders to buffers, micro triangular points to diamond tipped arrows, scallop-edged bars to giant, rough looking balls that look like they belong on a medieval mace. As the tool is powered, the interchangeable tip is rotated at high rpms. This rotating action scrapes matrix away and can get you close to the actual bone surface. Notice, I said CLOSE. You seldom want to employ a rotary grinder directly on the surface of a bone as this can grind straight into the bone's surface if you are not careful. Rotary grinders are not very good at removing large volumes of matrix, but are excellent for cutting away heavily glued sandy matrix, or working near (not directly on the surface of) very delicate fossils. The grinding, rotary action is usually less invasive than the vibrating needle tip of an air-scribe.

F. Sonic cleaners- To be honest with you, I have never tried to use one of these interesting contraptions. They consist of a small basin that preparators fill with water and a few drops of a cleaning solution. They are electric powered and once they are turned on, they vibrate at an incredible speed. They are generally used for cleaning jewelry and not fossils, but I have heard of other preparators using them with mixed results. I have one sitting on a high shelf in my shop, but I have yet to get up the courage to experiment with it. Anything that is small enough to fit into the basin is also probably too soft and fragile to be vibrated in that manner. Perhaps some shark teeth, ammonites or other common fossils might make for a safe experiment.

G. Micro-abrasion units- Micro-abrasion units were, less than 10 years ago, a highly controversial tool that many preparators mightily resisted. Today, they can be found in most large prep labs and used on fossils with a wide degree of preservation. A micro-abrasion unit consists of a tank that is filled with an abrasive powder and then pressurized with air. A small hose attached to the tank exits the unit from the rear and terminates in a little hand held "pen". When a foot pedal is activated, a tiny valve is released inside the main body and a high pressure jet of air and abrasive shoots out of the pen. Basically, the device is a miniature sand-blasting unit. The abrasive is aimed at the matrix by the preparator and it slowly or rapidly (depending on your air pressure) erodes the matrix away. There are several manufacturers of micro-abrasive units (also called "blasters" or "micro-blasters"), but the two most common used in fossil prep (micro-blasters have many industrial uses other than fossil prep) are made by *Comco* and *Crystal Mark*. Both companies make several different models for various applications. Each basic unit generally costs about $3000, so they can be quite expensive. Micro-abrasion is sometimes a dangerous thing in the wrong hands, hence many of the initial fears and misconceptions. It is generally a good idea to have some personal, hands-on training with the tool before blindly jumping in. If used correctly, they can really be used with delicate precision on both heavy and delicate fossils. If used incorrectly, they can blast right through bone surfaces and blast away loose fragments. You can minimize damage to the bone surfaces by adjusting the air pressure, experimenting with different sizes or types of abrasives and changing the tips on the pen. See Rule #141 for more tips on safely using this type of machine.

H. <u>Dust collectors and fume hoods</u>- These tools are essential when working with any micro-abrasive units or when doing chemical preparation. They remove fine dust particles, stray abrasive powder and chemical fumes keeping your lab partners safe from... well... death.

129. Keep Your Tools Clean, Organized and Within Reach: Just
like in the field, you want to keep all of your tools, glues, and equipment clean, organized and in reach when working with fossil vertebrates. You never know when you will need to reach for the glue bottle and activator. You're not an octopus- you only have two hands. If you need to hold two broken pieces of bone with one hand, you will need to quickly and carefully reach for the glue bottle with the other. You can't do that if your glue bottle is located clear on the other side of the room or the tip is clogged. When a surgeon needs a specific tool, he can't excuse himself from the operating room and go looking through the storage drawers for the right tool.

130. Switch Tools Often in the Lab:
Again, this rule is similar to rules for working with fossils in the field. Each tool has a specific use and a long list of positive applications and negative drawbacks. You will be

more efficient and produce better quality work if you use the right tool for the right task and switch between them as the tasks dictate. If you find yourself using nothing but a micro-abrasive unit, then you are more than likely not using the tool correctly. If you spend all day long using nothing but a pin vise, your prep time will triple and your productivity will decline. You might need to use a rotary grinder to get through a particularly nasty bit of glued matrix, but you will want to switch to a blaster once it's cleared. Likewise, sitting there all day with a blaster trying to remove a stubborn piece of rock is a waste of time and abrasive, if an air scribe would be better suited.

131. Expect the Unexpected: When you begin opening up plaster jackets and probing inside for their treasures, expect to find more than what is written on the jacket. Often, large jackets have bones contained within them that were not visible to the field crews. These include elements from the main specimen and microfossils. Also, expect to have other things show up that, based on the log sheet, should not be. Some bones appear to be solid and in good shape on their surface, in the field, only to show up as root-rotted, powdery, punky and delicate on the inside. If you expect the unexpected, you will be ready for that event.

132. Watch for Bite Marks, Healed Injuries and Other Pathologies: Preparators need to have a very keen eye. As you are removing matrix with whatever tool is most appropriate, be on the look-out for bone textures or structures that seem out of place. These abnormalities can provide you and other paleontologists with an exciting story about your animal's life and help unlock the secrets of its

behavior or environment. Unusual changes, to an otherwise normal bone, are known as pathologies. These can be anything from "bite marks", indicating predatory or scavenging activity, scar tissue, indicating disease or injury, bizarre structures that shouldn't be there, indicating mutations or genetic deformities, and many other most bizarre oddities. Upon discovery of these truly unique structures immediately stop preparation and consult with a supervisor, fellow preparator or nearby researcher. If confirmed, include a brief description of the pathology in your lab journal and consult with others on how you should go about preparing it. Depending upon the nature of the pathology, you may need to be extremely careful in how you approach it. You may just want to avoid it entirely. Be sure to take photographs of the pathology and show them to other academics.

133. Custom Build Your Own Blast Box and Design Your Own Prep Area When Possible: As a preparator you probably do not get to customize your area too much, but if you can, give it some thought and planning. You will need to have your area as utilitarian as possible. You will have to decide where to place air lines, where to store tools, what type of chair to use, height and size of a blast box, etc. etc. etc. Blast boxes are built (with arm holes) so that you can place your hands inside the box and work with fossil specimens, without getting dust, abrasive powder and flying chips of matrix in your eyes. They can be purchased from the same people who make the micro-abrasive units. Usually though, they are not designed for fossils specifically, and often those purchased from the manufacturer are very expensive. Custom designing a box is often better and cheaper than buying one. Use 1/8 inch – ¼ inch safety glass for the top and strong plywood for the

walls. Then caulk all seams. Determine how you want the dust collector to attach to it and how you want the specimens to enter and exit the box. Some people prefer to work on a high table, others a low one. Some like to organize all their tools in drawers underneath the box, others like them lined out on a metal tray. Some people are left handed, others right. Whatever the case may be, make sure you are comfortable and safe in your environment and that it suits you, your fossils and your style best.

134. Observe and Learn From Other Preparators: One of the best ways people can

learn is through observation. If you have the opportunity, step back from time to time and see how your fellow preparators are doing their job. They may have developed a particular style or technique that works best for them and might work best for you too. If you can, visit other preparation laboratories and see how they do what they do. It's not stealing top secret information if you emulate what you see in other labs. It's actually the highest form of flattery.

135. Label Everything as You Go and Consider Keeping a Preparation

Journal. As stated before, a journal is essential in the field. It is almost equally important in the lab. Keep track of specimens you have worked on and specimens you will work on next. Note the tools and techniques used during prep. Note any strange features or pathologies; keep a record of your time spent on individual projects. Make notes regarding any experimentation you try with glues or chemicals. This can help others determine if your tooth marks are actual tooth marks or accidental x-acto blade gouges. It can help scientist to interpret whether or not the bone surfaces were acid etched because they were stomach contents or due to careless blasting or chemical prep. Making notes helps you to remember where you were if you get called away from a specimen to do something else.

136. Keep a Record of Who Prepared What and How Long it Took Them: Also see rule #135. A record of who prepared what and how long it took helps supervisors to evaluate an individual preparator's progress and development. By knowing the time it took for bones to be prepared it helps supervisors to plan how long the remaining elements might take and through some quick math, how long the entire project might take. By noting who worked on a particular bone, should that bone wind up missing or damaged, you know who is accountable and have a lead in where to begin searching for answers.

137. Keep Track of Those Field Numbers: One of my biggest pet peeves in the lab is preparators who carelessly lose or mislabel field numbers. The field numbers are important because they keep track of where the individual element was found at the dig site, by whom and other key statistical data. If you lose the field number it is very easy to lose the data or create confusion within the data.

138. Your Technique Will Depend Upon the Fossil:
As a preparator your first job is to learn the techniques of others and then use them in key situations depending upon the nature of the fossil and the nature of the matrix. Once you have mastered those techniques and move on to more and more difficult projects, then you should feel free to try new techniques or experiments. You can never learn everything from an instructor. Each fossil is different. Over time you will learn new things and develop your own unique tricks of the trade.

139. Always Start with the Least Invasive Tool and Work Up:
This rule is self explanatory. Start with hand tools and see what can be removed with those. Then move up to mechanical preparation methods. If all else fails then try chemical. Don't use a blaster if a simple toothbrush or a few drops of water do the trick.

140. Keep Your Micro-Blaster Dirt Free:
A micro-abrasion unit is an expensive tool that if used correctly requires little maintenance. Occasionally, you like every preparator ever born, will get larger bits of sand or debris accidentally in the fill tank. When that happens the machine will become jammed and the particle will need to be removed manually. This cuts down on productivity and can make prep very frustrating. You can minimize dirt mixing with your abrasive, by not using long sleeved shirts while you prep (dirt and rock fragments often get stuck in shirt sleeves and fall out when getting new abrasive), by keeping your abrasive bucket or storage container closed tight and by not re-using abrasive

unless it has been re-processed and re-cycled (I still haven't found a place that will do this… there's a market out there!).

141. Watch Your PSI/Keep the Tip in Constant Motion:

As stated in the rules regarding tool use, the micro-blasting unit in the right hands can produce amazingly well prepared fossil specimens. In the wrong hands it can do unbelievable damage. There are several things that a preparator can do to make the use of a blaster safe. They include adjusting the air pressure (PSI- pounds per square inch), alternating the size of the pen's tip and experimenting with different abrasives. The first thing a preparator should do when first getting started on a specimen, is to experiment with the blaster on a piece of weathered float to see if the micro blaster should be used at all. Start with a pressure of around 60 PSI and slowly increase to a maximum of 120 PSI. If it seems safe, then you should experiment on more important pieces. If the fossil surface seems to be undamaged at 85 PSI, but then slowly begins to show signs of wear, then kept your PSI set to no more than 85. Generally, most fossil preparation is done within a range of 40 PSI to 120 PSI. Harder, denser, better preserved bones with thick skins are often fine at 90-120 PSI. Softer, more delicate pieces should be between 40-80 PSI. Once you dip below 40 PSI the tool begins to lose any effectiveness at all. If the fossil is so fragile that 60 or less PSI begins to damage it, try a different technique or adjust something else. Since every bone has some parts that are softer than other parts, you will need to adjust the pressure often, even on the same element. The next thing the preparator can do is to change the tips on the pen. There are a variety of different colored tips, each with a smaller or larger orifice. The tips with the larger orifice allow a wider stream and greater volume of abrasive to exit the tip of the pen. These tips are better suited to well-preserved bones with lots of attached matrix as the powder is focused over a larger area. The tips with a smaller orifice shoot out a narrow, low volume stream of abrasive. These tips are best suited for more delicate specimens where you want to target a smaller area. Another thing you can do is to experiment with the type of abrasive used in the machine. Abrasives come in a variety of different types and grit sizes and changing either can help. Some available abrasives include: Sodium bicarbonate, or baking soda (the most common type used), dolomite, glass beads, silicone dioxide powder, crushed walnut shells and the like. Each type has a different hardness (check the Moh's hardness scale), angularity and grain size. If the abrasive is harder than the matrix it will remove it easily. If the abrasive is harder than the minerals that have fossilized the bone- it will remove that too, easily. You want to use a type of abrasive that is harder than or equal to the matrix but softer than the bone. The last thing you can do is to adjust the distance between the tip of the pen and the surface of the fossil or change the angle in relation to the surface. Bringing the pen in closer will focus the abrasive on a finer point. Pulling it back will produce a "shot gun effect", where the abrasive hits a wider area, but does so with less force. A shotgun is pretty much worthless from over 50 meters. A blaster is pretty much worthless at a distance of two feet. The closer you get to the fossil's surface, the more power the blaster has. Altering the angle will also have positive or negative effects. However you adjust your blaster, you will always need to <u>keep the tip moving</u> to make sure you are doing little to no damage to the fossil.

142.Oil Your Pneumatics: All of your air
tools should get a few drops of oil at the start of each day. When working a full day or if you use a tool frequently during the day, add oil as needed. Air tools are also

expensive. Unnecessary damage to the tool and annoying time delays can occur if you fail to do proper daily maintenance.

143. Work (Prep) From What you Know to What you Don't:

This rule is easy. If you open a package and see a vertebra covered in matrix, but the left side has been pretty much cleaned off in the field, start with the left side. Follow the cleaned part around to the other areas where the bone disappears under the matrix. Work the edges backward from there. If you see an area of nice solid bone that soon disappears into a mixture of matrix, glue and fragments, start with the nice solid bone and work towards the ugly stuff. Never start preparing in the ugliest, most root rotted, zones. If you start there, you may end up damaging what would have been decent bone around the edges of the break, since you may not be able to see a clear distinction between the bone and the rock until it's too late. Take the easiest then move to the hardest.

144. When Prepping Fossil Bone, Start in One Area and Work Outwards from That:

For some reason many new preparators clean a fossil bone in a random and haphazard fashion. They clean off one end then jump around to the other. They start working an area and then give up for something easier. By the time these newbies are half way through, the fossil looks like a gnarly patchwork of prepped and unprepped areas. Whereas several of our field and lab rules tell us to work on the easiest spots first and then move to the harder ones (Rule #143), they do not say to chaotically bounce all over the damn place! When you do this, you run the risk of damaging areas that have already been cleaned. This is especially true when working with an air-abrasive unit, where stray grains of abrasive powder bounce off the matrix and strike previously prepared areas nearby. If the preparator is focused on where they are aiming the blast media, they may not realize that the stray powder is hitting another area and doing harm. The best method is to start in the easiest, most well understood area and work outwards from there in every direction. You work away from previously worked areas, not towards them. This way, if stray abrasive is hitting the bone somewhere other then where you are focused, it is hitting in an area that has not been prepared yet anyway.

145. 90-45-10: Depends on the Fossil and the Matrix: This rule is by no means hard and fast, for the preparator must be adaptable. The 90-45-10 rule is basically a guideline that tells you at what angle you and your tools should approach typical fossil vertebrates. When working with a hard, dense bone with excellent preservation and good, smooth, unbroken "bone skin" (surface texture), surrounded by hard dense matrix (like ironstone, highly compacted siltstone, claystone, or mudstone) you come at the specimen on a 90 degree angle, perpendicular to the bone skin. An air scribe at this angle can be used to "pop" the matrix off the surface clean with a little practice and a little luck. You may also want to approach at a 90 degree angle with a *Dremel* for the same type of bone with a sandier or less cemented rock, and a 90 degree angle from a far distance with a micro-blaster for very loose sand and silt. In areas where you are not sure of the exact depth of the bone surface or the skin isn't uniformly in a great shape, use a 45 degree angle with blasters, scribes or x-actos. When the surface is not so good or the matrix is really sandy, take a lower angle (0-10 degrees) to the bone surface with all your tools.

Remember though… each fossil is different and each matrix is different and you will need to be cautiously inventive to find the best angle of approach.

146. Hold Your Tools Like a Pencil:

I'm not entirely sure why, but some first –time preparators tend to hold their air-scribes, rotary grinders, dental picks and x-actos towards the back of the instrument. Sometimes, they wield their tools like one might a dagger instead of a precision piece of equipment. Doing this gives you little control over the tool and you are bound to slip. Hold your tools as close to the tip as possible, like a pencil, for the best control.

147. For Every Action There is an Equal and Opposite Reaction- Support the Specimen Correctly:

I once watched a paleo-technician hold a thin limb bone at one end with his fingers and then try to prep the distant, opposite end with a medium-sized air-scribe. This is beyond incorrect. Had I not stopped the preparator, the bone would have snapped in half and fallen to the floor! You always need to cradle your fossil specimens carefully, placing your hand directly underneath the area you are working on. If you are using a lot of force from an air-scribe you need to match that force with your hand underneath. When carrying your fossils across the lab, cradle them like you would a baby, with two hands fully supporting their weight. Preparators who carry fossil limb bones like they were track and field batons are often quickly escorted from the lab never to be seen or heard from again.

#147- After dozens of desperate pleas, Dr. Bonehead saw no other option to his nagging laboratory problem.

148. If Delicate, Think *Dremel* or Blaster:
Air scribes are powerful tools and can sometimes be dangerous around delicate bone. The vibrating tip can break thin, fragile specimens even if you are supporting them from underneath. When the specimen is small and fragile, switch tools to either a *Dremel* and/or a blaster. Of course, the smaller and more fragile the bones are, the better off you will be with simple hand tools. Even these, with the wrong pressure, can snap a bone in half. If you need additional power to remove a stubborn piece of matrix, a *Dremel* or a blaster are much safer for the specimen.

149. Use Your Pinky as A Guide:
After many years of prepping and field work, I realized that I often braced my right hand (the one holding the tool), up against the fossil using my pinky as a guide. It might be a little difficult to picture the grip with just words, but hold an air scribe like a pencil, near the tip, and then drop your pinky finger down to the surface of the bone, forming a semi-flat base. From this position use your thumb, pointing finger and middle finger to push the scribe in short jabs forward and backward. It may feel awkward at first, but I have found that this hand configuration is exceptionally stable. It prevents the preparator from accidentally slipping too far and gouging the bone surface with the air-scribe's tip.

150. The Right Glue For the Right Task:
As a general rule use thin glue for powdery, crushed and highly fragmented bone and use thicker glue for large breaks with well defined edges. Use thin, consolidant glue, like polyvinyl acetate (PVA), to coat the surface of your bones once you have finished cleaning them. This will soak into thin micro-fractures and pores

giving your fossil additional strength and a nice semi-gloss coat. It will offer limited protection from UV-rays and make it somewhat water resistant.

151. Try Not to Have too Many Projects Going on at the Same Time:

Whereas you never want to prep when you are frustrated and you do want to stop and work on something else until you relax, you do not want to have twenty partially prepared bones at your table at any given moment! You will tend to lose and confuse field numbers when you do this. You will lose track of your time on your prep logs. At some point, you will need to grit your teeth, calm your nerves, finish what you started and work through the tough spots.

152. Sometimes it's Best to Re-break and Re-set Fractures; Other Times Just Leave it as is:

Rule #152 is one of the most difficult things to teach young preparators and old preparators alike. The decision on when to re-set a fracture and when to leave well enough alone is a difficult one and varies from bone to bone and specimen to specimen. With time and experience and lots of trial and error you will learn when it's the best thing for the fossil and when it's the worst. As a general guideline, if the fossil is exceptionally thin, composed of multiple, irregular, weathered chips and chunks just leave it as is. If the break is large with a wide separation and the bone fragments are bigger pieces (dime sized or larger), and the edges of the fragments are jagged and/or easy to put back together again, go ahead and dissolve the glue holding them together, clean out the excess glue and dirt from within the break and re-set the fracture. Beware however, for once you begin taking bone apart piece by piece and trying to fit it all back together again, you may find that the pieces do not want to fit nicely in their original spots. Some bones behave plastically as they are broken and deformed under the ground. Pieces may not line up correctly or not at all. Once you start however, you have to finish, otherwise you will forget how the pieces were supposed to go back together. Allow for a lot of time to do this difficult task.

153.Frustration Leads to Destruction:

It is very important to get up from the preparation table once every hour to stretch out and relax for a moment. Fossil preparation can sometimes be mind-numbingly slow. Fossils can break at the slightest of error. This can all lead to frustration. If you feel yourself getting frustrated, get up, "put your hands in the air and step away from the fossil", before you loose your temper. Take a ten minute or fifteen minute break. Switch to another specimen for a little while. Whatever you do, don't prep angry. Don't prep angry!

154. You Can't Prep What You Can't

See: The glass on your blast box will get covered in abrasive powder over time. You will need to wipe it clean often. Both the inside of the glass and the outside. If you can't see what you are doing you will make mistakes. Also, it is a very good idea to wear a 2x or 10x, magnification visor to help you see the surface more clearly. 2x is generally sufficient for most large vertebrate specimens. The visor helps you to get in close to your fossils as you are preparing them outside of a blast box, and doubles as a pair of safety goggles. After all, if a stray fragment nails you in the eye, you definitely won't be

#153- AT FIRST, THE PREPARATORS ENJOYED THE INTERACTION WITH THE MUSEUM GUESTS THROUGH THE VIEWING WINDOWS. LATER ON, THEY REALIZED IT COULD BE A PROBLEM.

doing much prep for awhile. When working on smaller microfossils or very delicate skeletons you may need to use a high-powered microscope.

155. Photograph the Various Stages of Preparation: On key elements from important specimens be sure to take several photographs during the preparation phase. You would be surprised how useful these photographs are at a later date.

156. Chemical Preparation Should Be the Last Resort: Chemical preparation involves the use of acids to dissolve stubborn matrix around fossil invertebrates and fossil vertebrates. The trick is finding the right acid that will eat away the matrix, but leave the fossil unharmed. There are many types of acids and many of them have been tried in prep with varying results. They include acetic acid (vinegar), weak hydrochloric acid (HCL), oxalic acid, formic acid, sulfamic acid, and others both stronger and weaker. Some of these acids can be quite destructive to your fossils so be sure to experiment on a piece of matrix, concreted float or weathered fragments before pouring it on. Acetic acid works just fine for most carbonate rocks like limestone or rocks where the cement is a calcium carbonate. It will dissolve many fossils too if you leave it in the bath for too long. I have had mixed results with sulfamic and oxalic acids (CLR and other household cleaners) on iron carbonate concretions. Muriatic acid has been tried occasionally, but tends to be too strong for most fossils. A 2% HCL solution does the trick on many invertebrates but again test it first. Whichever acid you try to use, make sure it is diluted and weak. Don't expect to have much of a specimen left if you experiment with strong HCL or

sulphuric acid. Sometimes, all you need is good old common tap water. Most acid baths will increase their effectiveness if you keep the bath warm. Once you have soaked your fossil in your acid, in many cases you will need to soak it in a buffering solution to neutralize the acid, or the dry residue may continue to slowly eat away at it. No matter which acid you try or how you do it, don't expect chemical preparation to work overnight. Many specimens need to soak, then buffer, then soak again several times before the desired effect is realized. This can take weeks. If you try using chemicals to assist in prepping fossils, also be sure to wear the appropriate safety equipment and work in a well ventilated area. When in doubt, contact a specialist in chemical preparation techniques to make sure the acid you choose is safe to use in your circumstance.

157. Change the Filters on Your Dust Collectors: In a busy prep lab, your dust collectors will fill up fast. Remember to change them on a regular basis. If you are working in a lab where there are no dust collectors then I suggest you give OSHA a call. Breathing in a little sodium bicarbonate, dolomite, 100 million year old dirt, etc. will probably not kill you, but prolonged exposure with no access to proper ventilation is definitely not a good thing.

158. To Restore or Not to Restore That is The Question: Restoration involves repairing and sculpting missing pieces. This is done to help stabilize the specimen and for aesthetic purposes. Some techniques provide the specimen with additional strength. Others allow you to mold and cast a specimen safely. Some restoration, unfortunately, is done just to

improve how the specimen looks or to hide errors caused by poor field work or prep-work. The decision on whether or not you restore a specimen depends primarily on the scientific importance and stability of the specimen.

If you are working with an extremely important fossil you will want to keep restoration to a minimum. In fact, be sure to take lots of pictures of the specimen and show them to other professionals before you attempt any restoration whatsoever on extremely rare pieces. After this, you might want to do basic resto, by filling in any large fractures with either epoxy or apoxy putty. These can provide additional strength and stability across those large breaks. After all, you do not want the specimen to fall apart under its own weight, do you? Be sure not to cover any key structures or features. Most apoxy/epoxy putty's come in two parts that will need to be kneaded together with your fingers. This will begin the chemical reaction that will harden it. Start with small quantities of putty and use only as needed. Apoxy/Epoxy putty's are expensive and you do not want any waste. Their set up times vary from product to product, but you generally have about 10-15 minutes where the putty is workable. Try to get the putty into the fractures as deeply as possible and either sculpt it to lie parallel to the surface or leave it somewhat recessed below the surface.

If the specimen is going to be molded and cast, in order to prevent silicone or urethane from seeping into porous bone or broken fractures, you will also need to fill certain fractures with colored and textured water putty. If you try to mold and cast a specimen without it, you might have a mess on your hands as both silicone and urethane both will work their way into every crack and pore. Removing that excess will be a nightmare and can cause irreparable damage to the fossil. To do basic crack fill restoration, the water putty is mixed with pigment and pushed into large gaps with your fingers or clay working tools. You do not want to get any water putty on unbroken surfaces, so you will need to be constantly wiping the specimen with a damp cloth to remove excess. Once the specimen dries you will need to give it one last quick blast with an abrasive to remove any excess not removed by the damp cloth then use PVA to coat the entire bone once more to seal it. It is important to remember that basic crack fills with water putty does not provide much additional strength, so if you need to make it more stable, use apoxy putty first and make sure the apoxy putty is recessed far enough into the cracks that it can be covered with water putty and not be seen.

Ultimately, in my opinion, it is best to NOT do any large sculpting and restoration on original bone. I would recommend that you first make a mold and cast of the original (see Rule #160), without heavy restoration. Once you have a cast in your hands, then set the original away and you can sculpt all you want on the cast. Once you have sculpted on, textured, heated and bent the cast copy a second mold can be made from that. This method protects the original and provides you with two molds- one for research the other for mounting. It's more expensive and time consuming, but it can pay dividends in the long run.

If you absolutely need to sculpt missing structures on original bone to help stabilize the specimen or to create more complete casts, then it is best done with water putty (or even clay) which can be removed at a later date if necessary. Both epoxy and apoxy putty are far more difficult to remove even though they do sculpt easier and better. When you are sculpting missing pieces, make sure that the color of the restorative is slightly different than the color of the fossil. There must be a clear distinction between the two. This way, professionals can tell what has been restored and what hasn't.

Restoration for aesthetic purposes should only be done on specimens that have little or no scientific importance. A museum may consider aesthetic restoration if the specimen is going to be displayed and not studied in depth. Here, you would sculpt missing pieces and use a color and a texture that is similar to the original bone. If the sculpting is done well, it will be very difficult for anyone viewing the fossil to tell what has been restored and what hasn't. If this aesthetic restoration is for a museum mount or display, it is a good idea to have a permanent, written record of what has and what hasn't been restored, just in case someone comes along and wants to study it. In all cases, make absolutely sure the restoring technician knows about the fossil's anatomy. You do not want someone accidentally filling in natural foremen, fossa, pneumatopores or other natural openings. You also want to make sure they do not cover and obscure important pathological markings such as tooth marks and/or scar tissue. This would be bad. When in doubt do not restore it. If the intent of your restoration is to deceive someone about a fossils quality or condition, it will not make YOU look any better if you restore it.

159. Restoration Should Be Removable When Possible: Reversibility is one of the key things to good preparation and restoration. Ten, twenty or more years in the future, someone may come along with a new observation, technique or restoration material that will render your restoration obsolete. You want them to be able to remove the resto, if they choose, and do so in a way that it doesn't damage the fossil. Whenever you decide to make something "permanent" make sure you are doing it for the right reasons.

160. Molding and Casting? Entire books could be written on all of the wonderful tips tricks and techniques for molding and casting (the field could really use a good modern one- hint hint!). To include them here however, would have pushed us well over the 256 rules I started out with, so for space's sake, only a basic overview is given. Please visit the wonderful website,

www.fossilprep.org , attend an SVP preparator's section seminar or check out other books on the process for more details.

Basically, the process of molding and casting is how preparators and restorationists make exact copies (casts) of fossil specimens. The reason we make copies is not to try and deceive anyone, but for the original fossils protection. In the past, paleontologists would mount original bone skeletons, placing them in museum halls for all to see. The trouble was these skeletons weighed a ton. Due to the weight, large steel support structures often had to be built around them to keep them erect. In the past, this process was very invasive requiring technicians to sometimes drill right through the original bone in order to insert steel rods and pins that could then be welded together. The process was often damaging to the specimens and in many cases, pretty much permanent. Try removing a bone from one of these old mounts! Today when a technician mounts a real skeleton, they have ways to support the bones with brackets and steel holders, set on a removable steel frame, which makes drilling original bones unnecessary. Even today, however, original bone mounts are heavy, dangerous (you don't want a one-of-a-kind specimen hanging twenty feet in the air- especially in earthquake zones) and very expensive. It is much cheaper and easier to mount a plastic copy rather than original bone.

Casts can be made from a variety of different materials including plaster, plastic, expanding foam or other materials. The majority of casts are made from a liquid resin called *Por-a-kast*. This plastic is lightweight and non-toxic. The cast can be drilled for mounting, heated and bent (to remove original deformation), textured and painted (to match the original color of the fossil).

Before you can make a cast, you must first make a mold. Molds are made from a variety of things including plaster, silicone, urethane, latex, fiberglass and other materials. The following is a brief outline on how to make a simple two-part silicone mold. It is not intended to be a substitute for hands-on training.

A. Completely prepare your fossil bone and coat with PVA (polyvinyl acetate)

B. Restore (as needed) your fossil bone, making sure all major cracks have been filled with *Durham's* water putty or other restorative. For scientific specimens or research sets, do not sculpt missing pieces (You may consider making two molds- one directly from the un-restored original and a second from a cast with the missing pieces sculpted and textured- see rule #158.)

C. Use modeling clay to partially fill any soft spots, cracks, crevasses or holes (that you decided not to restore in part B) that you do not want the silicone to get into. Beware of silicone getting into your bone marrow. You will want to partially fill things like broken patches, cracks around teeth, pneumatopores, foramen and fossa.

D. Take a wooden board, large enough to hold the entire mold and wrap it in cellophane. Place the board flat on a work table. Place your fossil on the board as flat and stable as it will sit.

E. Next, try to imagine the bone's hypothetical split-line. A split-line is the imaginary line that will divide your mold into two semi-equal halves. Your split-line, will lie in the horizontal plane (not the vertical) and not have any major overhangs. (*Editors note: remember- the split line is an IMAGINARY line. Please do not physically cut the bone in half!*) The split-line is not necessarily along any bi-symmetric plane. Teeth, flat plate-like bones, digits and limb bones are all fairly easy to determine their split-line. You can find it by looking straight down onto the bone from above and finding the line on the bone that corresponds to its maximum outline. Once you have determined where the split line should go, try to keep that in your mind's eye or in some cases you might want to dash the line in on the bone with some sort of removable ink marker (test on a piece of float to make sure the marker is easily wiped clean before you try this on a completed bone).

Vertebrae and skull elements however, are more complicated (where a clear split-line without overhangs is harder to find) and finding the split-line takes practice to master. When you are making a two-part mold remember that one side of your split-line will be on one half of the mold and the other side of your split-line will be on the other side of the mold. If you have overhangs on one side or both sides of your split-line, you may have difficulty getting your original bone out of the mold (possibly damaging it in the process) once the silicone/urethane hardens. For these more complicated molds, your split-line will most likely vary in elevation dramatically. Using softened modeling clay (a heat gun, heat lamp, or microwave for a few seconds works fine to soften it), fill up the areas around the fossil bone to the edge of the split-line. Make sure the clay is placed completely around the fossil and there are no gaps between the bone and the clay. Use a metal clay working tool to "cut-in" a clean, straight line between the bone and the clay. Eventually, you will have a smooth clay base covering the bottom portion of your bone completely encircling it. The clay base should be relatively flat and smooth. The top of the bone should still be visible sticking out of the clay.

F.   Next, determine where you want the pour spout to be. This is going to be where you will pour your liquid *por-a-kast* into the mold. Make a wide cone shape out of a small piece of clay, by rolling it between your fingers. Then, place this pour spout on the end of the bone you feel is best. The plastic will also fill the pour-spout, so it will need to be cut and trimmed off later- keep that in mind when you determine where to place it. Blend the edges of the clay pour spout with the clay base so it becomes one.

G.   Take some more clay and roll it into long snake-like ropes. This is your "key". It will help the mold lock together and prevent leaks. Place the clay ropes around the bone onto the clay base, leaving at least ½ inch gap between the bone and the rope. Do not go over the pour spout. Using a clay working tool, blend the clay rope into the clay base.

H.   Trim off the excess clay on the base. Give at least ½ inch space between the rope and the edge of the base. Trim the pour spout so it has a flat top and conforms with the new edge of the base.

I.   Roll out more clay with a rolling pin until you have long flat strips of clay. Cut into a long rectangle (s). These are your walls for the mold. Stand them on edge and blend with the clay base so it becomes one piece.

J.   Write (in reverse mirror handwriting), the field number or bone type somewhere on the clay wall and on the clay base.

K.   Spray the inside of the mold and the top of the bone with a mold release. This will keep the silicone from sticking to the bone and the clay.

L.   Mix your silicone as per manufacturer's instructions. There are several different manufacturers; the one I frequently use is *GI-1000* from *Sterling Chemical*. The most important thing to remember about your silicone or urethane is the durometer scale. The lower the durometer the more flexible your mold will be. The higher the durometer the more rigid the mold will be. For fossil vertebrates you want a good balance. If the durometer is too high you will have difficulty getting the original and subsequent casts out of the mold. If the durometer is too low, your mold will tear more easily. Make sure your silicone is well mixed and try to get as many air bubbles out of it as you can.

There is no rush, as most silicones have long working times.

M. Pour the silicone into the mold, completely covering the bone to a depth of ¼ inch (min.) above the bones highest point. Set aside and let dry overnight.

N. When you return the next morning, your silicone should have (if mixed properly) set up and is now a flexible, blue, rubbery substance. Carefully pick the whole thing up (including the wooden base), and flip it over onto another wooden base. Remove the 1st wooden base. You should see the underside of your bone and the clay that you placed around it. Without removing the clay walls, remove the clay base and the clay ropes. Leave the clay pour spout in place. Make sure silicone did not leak below your split-line. If it did, trim off any excess to the split line.

O. Build up your clay walls as high as necessary to hold the next mix of silicone.

P. Spray a release agent into your mold, on all surfaces (bone, walls, clay, silicone-especially silicone… you do not want both sides of the silicone to bond together)

Q. Mix your silicone as in step N and pour into the mold. Set aside overnight.

R. The next day, carefully pull apart both sides of your silicone mold without damaging the original fossil. Remove the clay walls and clay pour spout. Take your original bone, clean off excess clay, and place on a clean storage shelf for scientists to come and study. Clean off any remaining clay from the mold, put both halves together and you are finished with your mold. Fill with *Por-a-kast* and in 15 minutes you have an exact copy to paint, bend, twist, build a skeletal mount and hang 20 feet in the air, to the "ooohhhhhs" and "ahhhhhhs" of your museum guests.

161. Use Databases Effectively to Accession all Fossils: Each significant fossil you collect in the field should be logged into a computer database. The majority of public and private museums, as well as universities, officially accession each one of their specimens into a computer database for tracking and research. Each specimen should be inventoried and all of its relative data, taken from the field journal, log sheets, lab journal and other research notes placed in this database. Database programs such as Microsoft Access or Past Perfect are often used. Once entered make sure you have a backup just in case. Institutions that also post this information on the web provide a great service for research.

162. Composite Specimens Should Be Clearly Noted: The majority of fossil vertebrate skeletons are nowhere near 100% complete. Most are less than 25%. As a result, some common skeletons hanging in the prestigious halls of museums and universities world wide are actually composites. A composite skeleton is a mounted skeleton (cast or original) made from the bones, or casts of bones, of two or more individuals of a similar size. Most of the Triceratops skeletons that are in museums today are in fact, composite skeletons.

#162- DESPITE DR. BONEHEAD'S BEST ATTEMPTS, FEW PEOPLE BELIEVED THE BONES WERE FROM A SINGLE INDIVIDUAL."

Composite skeletons are what you might call a necessary evil. They provide museums with a wonderful display piece. They help promote the science and help to recruit enthusiasts and future professionals. So long as the specimen is clearly noted as being a composite there are no problems. The problems occur when the specimens are not listed as composites and this makes their use in science problematic. Especially when the records of which element came from which specimen has also been lost. If you are compositing a specimen, make sure you keep a record of which elements came from which specimen. When there is a great deal of restoration or reconstruction/sculpting this also needs to be clearly marked. Too many mounts in museums do a poor job of explaining this to the public leading to confusion and bad PR. Make sure those guests are also aware that the specimen hanging in your hall is a cast and not an original. Most will understand why it's a cast if you explain it to them. Only a handful will ask for their money back.

163. Register Your Fossils and Images with the AAPS online Registry or Other online Registries: If you have a privately held specimen of any significance it is a good idea to register it with a reputable group. Make sure you list as much historical and scientific information about the specimen as possible. Allow the specimen to be studied by researchers who request a viewing.

164. If You Sell, Sell Responsibly: If you are engaged in commercial paleontology try to make sure the specimens you find get into a good home. There are three main types of vertebrate fossils; 1) Those with <u>scientific significance,</u> which includes any fossil

(vertebrate, invertebrate, micro, trace, plant, ichno, etc.) which appears to represent a unique taxonomic, behavioral, biological, pathological, stratigraphic, geographic or preservational setting that is not represented by more than approximately 100 well documented specimens held in the public trust or generally considered to exist, 2) Those with commercial significance, which include disarticulated, isolated, incomplete elements from well known genera, or partial skeletons from common genera that have little scientific significance, but are prized by collectors, educators and enthusiasts, and 3) Those considered common fossils- those that have little or no scientific interest due to poor quality, poor condition, general abundance and are from very well known genera from a large geographic range. Responsible commercial groups would not sell a scientifically significant specimen to just anyone in off the street. They need to make sure these important fossils find their way into a reputable repository, with free and open access, where they can be studied by anyone interested in doing so. If you do collect a specimen that is scientifically significant, please try to either donate or sell that specimen only to a professional repository (public or private institutions where they can be freely and openly examined by other scientists). If, like *Indiana Jones* said in the movies, "It belongs in a museum!", then try VERY hard to get it there.

165. Know What You Have Before You Sell: If you do decide to sell fossils, it might be a good idea to take a good look at them, BEFORE you decide to sell them. As a responsible person, you need to decide if the specimen has scientific interest. You can't do that with poor research and observation. Not only does this not help the science of paleontology, but it is also bad for business. I know of a few cases where a commercial collector found something, then sold it, only to realize that

there was more of it in the ground. There have also been dozens of cases where the specimen was originally misidentified and sold under that misidentification sometimes to the benefit of, but usually to the detriment of the buyer.

V. Rules For Research and Publishing

166. Publish Frequently: The best, most well-known, most respected paleontologists in the field publish as much as possible. Some, once every few months. How "the best" find time to do this, I have no idea. I barely have enough time to turn off the alarm clock, walk the dog, respond to e-mail and eat three meals a day before *American Idol* starts! How ever you find the time to do it- do it. If you do not publish you will slowly fade in this field. There are many avenues to publish your research in print and online scientific journals, academic presses, popular science magazines and the like. A great list of all the journals that publish paleontological topics can be found at http://cactus.dixie.edu/jharris/Journal_Links.html (thanks to J. Harris). Remember, the more you publish, the more your name gets out. When that happens, certain doors that were previously closed will start to open.

167. There's Always Something to Be Researched and Published: Too many young paleontologists find themselves suffering from writer's block. They really want to publish on SOMETHING; they just can't figure out what that something should be. Often these young paleontologists come up with great research ideas, but end up talking themselves out of it because they suddenly get stage fright, tell themselves they are not qualified enough to write about this or that or worry that a particular topic has been done to death. Sometimes they are steered along a path that is really their graduate advisor's passion and not

their own. For those looking for research ideas, I have compiled a short [sic.] list of suggestions. Just fill in the blanks for your own custom project (and no, you shouldn't play *Mad Libs* with the blanks! Well okay, I guess that might be fun after all!):

A. Taxonomy:
  1. A Cladistical analysis of the _____ family
  2. A Phylogenetic Analysis of the Sub-Family _____
  3. A Re-evaluation of the Genus _____
  4. A New Classification Scheme for the _____ type _____
  5. Preliminary Identification of _____
  6. A Description of a New _____ from the _____ Formation
  7. Comparison of the Members of the _____ family
  8. Why _____ is a Valid Genus
  9. Why _____ is an Invalid Genus
  10. Fossil _____ (s) from the _____ Formation
  11. _____ , a Well-Preserved _____ from the Country of _____
  12. Fossil Vertebrates from Northern _____
  13. _____ From the Upper Paleocene
  14. The Benefits of Cladistical Analysis
  15. The Problems and Errors in Cladistical Analysis
  16. New Stem-Lineages of the _____

17. How the Relatives of the _____ Family are Not Related to the

_____

18. The Morphology and Systematics of the _____ Order.
19. Node Based _____ Lineages
20. What Needs to be Done with _____ Family Tree

B. Physiology and Functional Morphology:
1. Dinosaurs are Warm Blooded Because of _____
2. Dinosaurs are Cold Blooded Because of _____
3. Dinosaurs were Neither Warm Blooded nor Cold Blooded and were actually _____ Based Upon

_____

4. How To Estimate the Metabolic Rates of _____
5. Estimated Metabolic Rates Compared Between Extant _____ and Extinct

_____

6. Bone Histology Analysis of

_____

7. Lung Function in Avian

_____

8. Cranial Anatomy of Cretaceous

_____

9. Soft Tissue Reconstruction of

_____

10. Muscle Mass and Weight Estimates of _____ from _____
11. Variations in the Tooth Count of _____ and its Relation to _____ Strategy
12. The Neck Function of

_____;

13. An Analysis of Limb Proportions of the _____ Family.
14. Growth Rates Among

 _____

15. The Function of the
 _____ and it's Relation to the _____
16. Reconstructing the _____ Muscles in the Genus _____
17. How Much Food Did a
 _____ Need to Consume; A New Mathematical Model
18. Stomach Contents of an Early Oligocene _____
19. Swimming Motion of
 _____ Modeled by

 _____

20. Mechanics of the
 _____ Wing

C. Evolution:
 1. The Timing and Mode of Evolution
 2. Punctuated Equilibrium is Valid Based Upon _____
 3. Why _____ Proves Punctuated Equilibrium is Invalid
 4. Phylletic Gradualism shown in the Last _____ Years of the

 _____

 5. Allopatric Speciation of Late Paleozoic _____
 6. Tracing Family Trees Through Time
 7. Bottlenecks in the History of Life
 8. The Adaptive Landscape of

 _____

 9. Adaptations in the
 _____ Family
 10. Mutations and the Fossil Record.

11. The Signor Lipps Effect and its Role in the Fossil Record
12. The Genetic Barrier of _____ and the Divergence of _____ and _____
13. Sexual Section and the _____ Families Success
14. Allometric Growth and the Species _____
15. Ontological Variation within the Clade _____
16. The Late _____ Arms Race
17. Extinction and Evolution; The Death of _____ and the Rise of _____

D. Extinction:
1. The Periodicity of Extinctions Validated by _____
2. The Myth of Periodic Extinctions as Evidenced by_____
3. An Overview of _____ (name extinction event) Survivors
4. The _____ Killed the Dinosaurs
5. The Myth of Mass Extinctions
6. Survival Strategies of Early _____ _____
7. Geophysical Consequences of Large Impacts
8. Atlas of Large Impacts
9. The _____ Impact from _____ and it's Relation to _____
10. A Two-Fisted Extinction Trigger- Impact and Antipode

11. Climatological Records From
_____ as Evidence of
Permian Climate Change
12. Early _____Volcanic
Activity Compared to the
_____ Eruption on
_____ in Modern
Times.
13. How and Why _____
Goes Extinct at the End of the
_____

E. Behavior:
1. Nesting Behavior in
_____ _____

2. The Evidence for Parental Care in
_____

3. Predator or Scavenger; The
Magnificent or Not So
_____

4. Pack Hunting Strategies in the Family
_____

5. Onshore Egg Laying as Evidenced by
_____ for the Genus
_____

6. The Role of _____ in
Attracting Mates
7. Juvenile _____ Behavior
8. Pathological Evidence From
Specimen _____ Showing
_____,
_____,
_____

9. Extinct Behaviors of
_____ Compared to
Extant Behaviors in the Modern
_____

10. Hunting Strategies of
_____

11. Herbivorous Activity of
    _____

12. The Social Pecking order of the
    Genus _____

13. Herding in the Clade
    _____ as evidenced by
    _____

14. Cannibalism in the Species
    _____ shown in the
    Specimen _____

15. Hibernation in
    _____

16. Interspecies Rivalries Between
    _____ and
    _____

17. Intra Species Competitive Strategies
    of the Genus _____

18. Love in the Past; Mating in the
    _____

19. Predator Traps; an Analysis of the
    _____ Site

20. Stalking (or Gregarious, or Herding,
    or_____) Behavior as Evidenced by
    the _____ Track
    way Site.

F. Geology, Taphonomy and Geo-
   Chemistry:
   1. Ecological Setting of a Middle
      _____ Bone Bed
   2. A Cross Section Through the
      Pleistocene _____
   3. The Stratigraphy of
      _____
   4. Correlating Units Across the
      Continent of _____
   5. Age Dating Analysis of the
      _____ Site

6. Pollen Analysis of the
   _____ Site
7. Petrographic Markers for

   _____
   Paleonenvironmental Analysis
8. The Use of Subsurface Structures to
   Estimate the Fossiliferous Extent of
   the _____ Formation
9. Known Fossiliferous Zones of
   _____ Formation
10. The _____ Formation
    and its _____ Fossils
11. Paleonenvironmental Analysis of the
    _____ Site
12. _____ Atlas of
    Known Fossil Localities
13. Biogenic Concretionary Masses; How
    they Form and Why
14. _____ Concretions in the
    _____ Formation and their
    Use as Paleontological Marker
    Horizons.
15. Plate Tectonics and Continental Drift;
    Bringing the Genus _____
    and _____ back
    Together Again
16. The Oldest Record of
    _____ in

    _____
17. Mass Death Assemblages in the
    Genus _____ a Re-analysis
18. A Taphonomic Assessment of the
    _____ Site
19. Paleosol Horizons in the
    _____ Formation
    and Their Paleontological
    _____ for

    _____

G. History
   1. The Life and Times of
      _____

   2. The Paleontological and Geological
      Contributions of _____
   3. The _____ of the
      Great Bone Wars
   4. The Field Journals of
      _____: What His/Her
      Story Can Tell Us.
   5. Paleontology in the Ancient
      _____ World
   6. Paleontology in Myth and Legend
   7. Paleontologists of the 19th (or 20th, or
      21st) Century
   8. The Paleontology of
      _____ (country, state,
      province, county, town)
   9. Paleontology in Modern Culture
   10. A History of Fossil Collecting Laws
   11. The Politics of Paleontology
   12. Paleontology in the Movies
   13. Modern Controversies in
       Paleontology
   14. A History of Paleontology's Greatest
       Achievements
   15. A History of Paleontology's Greatest
       Failures
   16. _____ Culture and
       Paleontology
   17. Fossils as Art
   18. A Review of nomia dubia from the
       19th Century
   19. The _____ (good,
       bad, indifferent) Impact of
       Commercial Fossil Collecting based
       upon _____

    20. How We Thought They Were; The
        Concepts of _____ Over
        Time.

H. Fossil Preparation:
    1. Micro-Preparation Techniques on a
        _____ skeleton from
        _____
    2. How to Remove _____ from the
        surface of _____
    3. Selecting the Right _____
        from Manufacturers
    4. Chemical Preparation Techniques
    5. The Use of _____ (chemical) in
        Preparing a _____ skeleton.
    6. Paleontology and New Technology

I. Fossil Exploration:
    1. Using Aerial Photographs to Locate
        _____
    2. Geophysical Tools and their Use for
        Fossil Exploration
    3. Exploring with _____
    4. Hi-Tech Paleo
    5. A Ground Penetrating Radar Survey
        Over a Known Eocene _____ Bone
        Bed
    6. Using Index Fossils in Paleontology
    7. Tracing Paleontological Markers
        along the _____ Mountains.
    8. Exploring the _____ Desert
        (or Mountains, or Coastal Plains, etc.)
    9. Diving For Fossils in a Marine
        Environment

J. Field Work:

K. Mold Making and Casting:

168. Never Stop Asking Questions:

Some of the best science is achieved by questioning the results and conclusions of other researchers. The more questions you ask the more thorough your analysis of a particular topic, becomes. As you read the research of others look for things that might prove or disprove their hypothesis. As you do, consider your own experiences and observations and determine how those can bolster or refute their data. It's a great way to come up with research projects of your own.

169. Organize Those Files: As you are doing

your research, try to organize your stacks of scientific papers into an easily reachable reference system. Group your data so you know where to find it when a question comes up. Place like papers in like computer files. Organize your books and journals. Place photographs and other images in handy locations. Sort out frequently used papers, photos, documents, and charts so that you always know their locations. Time is of the essence. When you need an answer to a question its better to have it nearby then buried in a disorganized pile of obscure reference papers.

170. Join Professional Organizations:

Joining professional organizations is a great way to build business connections, make friends and learn the art of research. Most organizations have annual conferences that have research presentations, seminars, poster-sessions, field trips, and guest lecturers. You must attend as many of these as possible. Pick and choose the talks that are of the most interest to you and carefully learn from the presenters. Take notes, ask questions. Try to meet with the lecturers during breaks. Eventually, with a little luck and lots of hard work, in a few years you will be the one standing on stage and giving that 20 minute-long power-point presentation. Organizations include The Society of Vertebrate Paleontology (SVP), The Geologic Society of America (GSA), The Paleontological Society, The Association of Applied Paleontological Sciences (AAPS), Society for Sedimentary Geology (SEPM), The Western Interior Paleontological Association (WIPPS), The Mid-America Paleontological Society (MAPS), and many other state and regional organizations and fossil clubs. Many of these organizations sponsor their own journal where you may one day publish. Joining any of these can aid you on your quest.

171. Effectively Utilize the Internet:

The internet is a wonderful resource for all of your research needs. Knowing how to safely and effectively work the internet can save hours of time in a futile search. When using a search engine enclose your search subject in quotation marks to cut out a lot of chatter. Adding "PDF" to a search subject will often bring back many online scientific papers that you can download free of charge (good advice from Ken Carpenter). Join free and paid websites that support paleontology, geology or biology. Read, copy and file important bibliographies on specific topics related to your research. List frequently visited sites with the most useful reference information on the top of your favorites list. Search *Google Images* to find handy, sometimes royalty-free charts, graphs, photographs, artwork and even links to interesting sites you didn't know existed. Join Paleontological mailing lists and online chat-forums. All of these things can help you find information and conduct your research.

172. Visit the Nearest University Library: Whereas some research papers can be

found online, most will only be found in university libraries. The majority of university libraries have shelves and shelves of professional journals that have much of the data you will need to help you analyze your own fossil discoveries. Use the bibliographies found online to help you quickly find all of the publications you will need. When you find an article, you may simply want to photocopy it and look for the next one. Reading them, then and there, will take time. If you photocopy your articles you can take them home and read them at your leisure. Plus, you will grow your handy references and personal library.

173. Visit the Local Museum: When doing research be sure to visit your local museum. You never know what resources they may have available to you that might help in your study. You will certainly want to visit museums to view specimens first hand. No one likes a paleontologist who writes research papers about specimens they haven't even seen up close. Make an appointment with the museum's curator several weeks in advance of your planned trip. See if they can help you to find the information that you are looking for. Most will be happy to help legitimate researchers.

174. Specialists Are More Prone To Extinction: Just as animals that specialize are more prone to extinction, so too are paleontologists that concentrate solely on one aspect of research. If you specialize only on the limb structures of Pleistocene rodents from Europe, expect to only be called upon when a question arises about the limb structures of Pleistocene rodents from Europe. How often does that happen? You'll wind up a half crazy person sitting in a basement, measuring tiny rat legs, waiting for *National Geographic*

to call. Don't limit yourself this way. It's great to be an expert on one thing, but don't let it be your master. Don't let the thing you study or the theory you champion become the research embodiment of who YOU are. You need to be more than just that to survive and prosper in this game. Don't just be an expert on Pleistocene rodents- be an expert on all Cenozoic fossil rodents. Don't be an expert on just Cenozoic rodents- be an expert on all rodents or better yet all insectivores. Don't just be an expert on all insectivores- be an expert on mammals in general. Don't just know mammals; know a little bit about reptiles and fish, and evolution and whatever. Be a Jeffersonian. If you broaden and diversify your research interests, you will have a better shot at advancing in the ranks of paleontology. You will expand your pool of research options. You will be more open-minded to working with peers and using a multi-disciplinary approach to solving problems. You will tend to break Rule #179 more frequently if you are a specialist. If you do not diversify, something may come up that will render your entire life's work invalid or worse yet, make your specialty obsolete or insignificant and you will wind up going extinct yourself.

175. Use a Multidisciplinary Approach:

When you are doing your research and developing your hypotheses occasionally stop and think if other sub-disciplines might help you to attain supporting arguments for your conclusion or.... being fair to rule #175, to refute it. For example, let's say you are working on a paper involving the behavior of pack hunting in Dromaeosaurs. You have been using a particular dig site that has taphonomic evidence to support your hypothesis. Consider including supporting evidence gathered from modern day animals (biology) or other closely related species. Consider looking at the structures of the hind limb and forearm, the teeth and the skull to determine the biomechanical capabilities of the raptors in question (biology and physics combined). Let's say you are working on the Permian-Triassic Extinction. You were just looking at one aspect of it, a biological perspective, but when researching you stopped thanks to rule #175, and began to consider data published from people who looked at the extinction using different methods such as climatology records or geo-chemical analyses. You decided to look into geologic records involving plate tectonics, volcanics and stratigraphy. You considered some research by an astrophysicist friend of yours, working on something completely different, but it linked up with what you were doing. You factored all of these sub-disciplines together and it gave you an idea that hadn't been thought of before. Some of the best research papers, the ones with the most lasting, widespread use, use a holistic, wide-world, multidisciplinary approach.

176. Understand The Terminology: As you are researching you will occasionally come across a bit of terminology that you do not recognize or understand. Whenever you see such terms, you must stop and try to determine exactly what it is the author is talking about. Don't skip that step and assume by context clues you've figured out the meaning. Depending on the author that can sometimes be difficult. Make every attempt to grasp the concept in question before moving on. If you don't, you may end up confusing yourself and may develop incorrect usage that you will perpetuate in your future work.

177. Collaboration is Good: Sometimes it's best to work alone. Sometimes it's better to work with other people. If you are following Rule #176 and taking a multi-disciplinary approach, you will want to consult with

your peers. In the process of doing this you may find that you can produce a better-rounded paper/presentation if you join forces with them. Differentiate tasks, experiments, research or sections of the paper. Put one person in charge of one section and another on another section. When you pool your talents many great things can be achieved.

178. Contact Local Professionals and Experts:
Once you have completed a research paper you will need to have it peer reviewed. Call around and ask if your fellow paleo-geeks have a few hours to look over your data and conclusions and give your paper a good, hard critique. If your paper is on something that is specimen specific, invite them to come by for a private viewing so they can see the specimen up close. Get your paper in the hands of some experts (ones you can trust) to get their take on your research. If they are being honest, they will be able to point out potential flaws in your reasoning or problems with your hypothesis that you didn't consider. You can then use these constructive criticisms to make the paper stronger. You can either go back to the drawing board, continue to research more thoroughly to counter the arguments or toss the paper in the trash and move onto another topic that is completely unrelated. So long as your friendly neighborhood experts were being honest, no matter what they say (positive or negative), their advice will help you to make the right decisions.

179. Listen to the Opinions of Your Peers:
Many young paleontologists ignore or don't even bother to ask the advice of experts before they publish. Sometimes this is out of pride. Sometimes out of fear. Some are afraid that reviewers might not be honest

with their critique or even steal their research ideas. Listen to the advice you are given and then do the best you can with it.

180.Get a Second Opinion: Whereas Rule #179 tells you to listen and take into consideration what your peers, friends and experts say, Rule #180 tells you to not just follow them blindly. If you truly believe your paper has merit and is well supported, then by all means seek out second, third or twentieth opinions. If you get to your 20th reviewer and you are still hearing the same things, then you better take heed and change course. If not, and you are that pig-headed, then expect to be raked over the coals by readers that are not so close to you. They won't be as kind.

181. Research Opposing Viewpoints and Evidence: Okay, so you have your research idea and an outline of how you are going to proceed with gathering data, developing arguments and conducting experiments to test your hypotheses. That's great. Perhaps you have already run your idea and observations past noted experts in the field and gotten their opinion of your project. Before you finish developing your arguments however, there is one other thing you must do; study other papers that have come to a different conclusion than yourself. Your arguments will be stronger if you have looked in great detail at other papers which disagree with your conclusions. For example, if your project is dealing with catastrophic extinction events then not only should you research papers which support catastrophic extinction events and support your own data, but also papers which provide evidence for gradualistic extinctions that might refute your data. By doing this, you can find holes in your arguments that need to be filled before you publish. You

will know where the criticisms of your research are most likely to come from and upon what basis. You should, if you have analyzed the opposing views carefully enough, be able to counter their arguments with evidence of your own.

182. The Easiest Answer is Probably the Correct One: This rule, also known as Ockham's Razor or the Law of Parsimony, tells us that in any research question, where there are multiple, competing hypotheses, the hypothesis which requires the least complicated process and fewest assumptions, is probably the correct one. For example, let's say we have a dig site where there are bones with bite marks and lots of shed carnivore teeth. The teeth could very well have been washed into the site, but they appear to all be from the same size and species of carnivore. You find no other species teeth or microfossils at the site. The easiest answer would be that the shed teeth were from the same carnivore and that the tooth marks on the bones were caused by that carnivore feeding on the carcass. If the bite marks and the teeth match you have strong evidence for this. As stated before, the teeth could have been washed in, but in this example, you would need to somehow explain how they were sorted to species and size, and why no other microfossils were found at the site. If we reverse this, and say that we have found teeth from multiple carnivore species of various sizes including lots of microfossils, and that the bite marks do not match up with the majority of the teeth, then we have a different story. It is far easier to explain the presence of microfossils and carnivore teeth by them being washed into the site by stream transport than to have multiple carnivores also feeding at the site, creating bite marks that just don't match up.

This rule is frequently used in paleontology when working on the taxonomic relationships of various genera. Each genus has its own set of characteristics that set it apart from other closely related taxa. Some of those characters are shared and others are unique. The shared characters help to define all the common traits that the entire family of animals have. The unique or derived characters tell us whether the animal is more advanced or more basal. The number of steps or changes that need to be made to get from one character to the next can be great or can be few. When constructing family trees, the most parsimonious tree, and the tree that is probably closest to the truth, is the one which requires the fewest steps from one character change to the next.

183. The Easiest Answer is NOT ALWAYS the Correct One: Whereas

Ockhams razor is accurate 97.9% of the time and is USUALLY the correct answer, the modifier of "usually" indicates that the easiest answer is NOT ALWAYS the correct one. Sometimes life was more complicated than simplified, sterile answers will allow. Science too often tries to pigeon-hole natural processes into convenient little boxes that are cut and dry. Life, however, isn't always cut and dry! Sometimes Ockham's tendency towards over-simplification fails to take into account multiple variables and alternative hypotheses that are equally plausible. For example, in any extinction event, the easiest answer is a single cause or a single trigger. Traditionally, this is how scientists have approached the study of extinction events. It is far harder to comprehend (and for that matter, provide evidence for) a multi-causal extinction because it is more difficult to analyze the complex variables involved in such a study. Let's say we have evidence for a devastating impact at the K-T boundary. We could say this was the trigger, but other things might also be in play, that combined, was the real cause of the extinction. The impact may have triggered massive volcanism at its antipode. There could be climatic variables that were occurring prior to it that played a significant role. There could be hundreds of factors involved. That's complex, but equally plausible. Not only is it plausible, but, all things being equal, it may well be what actually occurred. Remember this when analyzing any natural system (including taxonomic assignments based upon the parsimony of cladistic trees) and take into consideration alternate, more complicated conclusions for the same data.

184. Once You Have Eliminated the Impossible, Whatever Remains No Matter How Improbable, Must Be the Truth... (Ehem) Best Answer at the Time:
Thank you Sherlock! Sir Arthur Conan Doyle's famous detective offers us a good deal of wisdom in the above rule. A great deal of paleontological detective work is not in finding precise answers, but rather eliminating the impossible hypotheses which definitely do not fit the observations. Once you have eliminated those impossible hypotheses what remains are our best guesses as to what is correct. It doesn't mean that the final conclusion is indeed the "truth", but the best, most logical guess at the time.

Rule #183 and #184 also stress that sometimes life throws us a curveball or two. Some fossils, skeletons, dig sites or rock formations just don't seem to make sense to us at first. As we start coming up with possible explanations for our observations we need to take into account every possible explanation conceivable no matter how silly or seemingly illogical. Then as more evidence and more observations are made we eliminate those which clearly do not fit.

185. If You Cannot Defend Your Position You Haven't Got One:
Paleontologists simply love to argue. Most of them are

well read, well trained and can support their wild hypotheses with strong evidence. Some are better verbal combatants, others are better in the written arena. No matter how it is done, they each, in their own way, try to defend their positions (some logically and some illogically) and in doing so, establish ones (positions). You will also need to learn how to defend your positions in verbal, visual and written ways. If you are going to argue a position on any paleontological topic you had better have sound, logical evidence to back it up. You will need to know why you believe the evidence supports your position. You will need data and facts and figures and statistics as your weapons. Only when you have this as your knowledge base, can you even think about claiming one position or the next.

186. Absence of Evidence Is Not Evidence of Absence: Just because you haven't found direct evidence to support a hypothesis, doesn't mean that the evidence does not exist somewhere. It does not prove your hypothesis is invalid. The fossil record is full of holes. It is highly imperfect. Most of the organisms that once called this planet home left no trace (in the fossil record) of their existence. Just because you do not find specimens of them, is not proof that they didn't exist. Just because we have not found certain "missing links", doesn't make evolution invalid. There has never been a complete dromaeosaur skeleton found in the latest Cretaceous of North America. Does this mean that there were no raptors running about in the last years of the dinosaurs? No, certainly not, for we find their isolated teeth and isolated bones all the time. In 1980 when the Alvarez group was publishing their end Cretaceous impact theory, they did so with limited evidence. They did not have an impact crater as a smoking gun. It wasn't until 11 years later in 1991 that the Chicxulub crater was discovered and correlated to the K-T extinction event. The

point is, do not fall into the logical fallacy of "argument from ignorance". Do not assume an argument is invalid simply because it currently lacks evidence.

187. There are Big Differences Between "Facts" and "Hypotheses": Okay, let's define our definitions shall we? A fact is defined by the Oxford American Dictionary as "something known to have happened, to be true, or to exist". It is absolutely true. It is THE truth. It is devoid of opinion and subjective observation. It simply is. A hypothesis however, is different. It is defined as, "a supposition or conjecture put forward to account for certain facts and used as a basis for further investigation by which it may be proved or disproved.". It is a best, reasonable, logical guess based upon the available data. A hypothesis then is based upon certain facts or observations, but not something that is conclusive by itself. Unfortunately, many young scientists and paleo-professionals often confuse facts and hypotheses and use them interchangeably and incorrectly in their work. This can lead to a great deal of confusion. For example, if you say, "*Pachycephalosaurus* butted heads with its rivals", then you have stated this (presented it in such a manner as to be interpreted) to be a FACT. This is wrong. It is not a fact that Pachycephalosaurs butted heads with their rivals; it is one particular hypothesis of many. In the absence of a time machine or some other way to concretely prove this, it is merely a hypothesis with good supporting evidence. It is one possible hypothesis out of many, used to explain the fact that most Pachycephalosaurs have a thickened, dome-like cranium shaped like a football helmet. The above statement in the hands of a responsible professional should be presented as "The evidence suggests that Pachycephalosaurus may have butted heads with its rivals." (or something similar to this). As you are reading the work of others or watching any television program on

ancient life watch out for these hypotheses masquerading as facts. Do not be fooled by them. You will find that they are prevalent throughout the science. Everything from bold statements like "Global warming is a fact that can not be denied", to subtle statements like "The warm blooded dinosaurs had to eat large quantities of food", confuse hypothesis with fact. Sometimes it's done by accident. Other times, it's done with the intent of advocating a particular position and gaining popular support.

188. Define Your Definitions: When writing a scientific paper, you will need to define certain things to your readers. Many of these things may seem obvious to you, but to others ummmm.... not so much. If you use a bit of obscure terminology in your paper, then it's a good idea to include a quick definition of its meaning. Certainly, you can not do this for every term. You must assume that your readers will know a little bit about paleo-speak before they read, but try to make it as easy as possible for them. You want to make sure that your readers understand your arguments and are not befuddled by your vocabulary. You want to also make sure that your definition of a particular concept is understood by your reader, who might have a different definition of the same concept. What is "sudden" in geologic terms to one author may mean 500,000 years to 5 million years. To another author it may imply 100 years to damn near overnight. Make sure you tell your readers what you mean when you say "sudden".

189. Beware of Subjective Descriptions:
In many taxonomic descriptions of families, genera or species, some authors will use subjective terminology to describe what they are seeing. The trouble is that subjective description often varies from author to author. What is "robust" to one author is "massive" to another.

What is "gracile" to one author is "delicate" to another. This can lead to confusion and misinterpretation. Subjective description is fine, so long as you define what you mean in more quantitative ways as well. Beware of descriptions like: strongly curved, slight, deeply set, shallow, box-like, triangular, irregular, well rounded, pronounced, closely striated, coarse, dense, plate-like, etc. etc. etc. without any further explanation.

Another problem with subjective descriptions is in the comparisons of taxa. Often, an author, when trying to argue for a separate status of a specimen, will compare it with another closely related taxa. The comparison usually involves saying things like; "The delto-pectoral crest is longer and more broad than "X" genera but shorter and wider than "Y" genera. If, the reader has a specimen of "X" genera and "Y" genera right in front of them to compare it to, then its not that big of a deal. In most cases however, your reader will not have either "X" genera or "Y" genera present when they are reading your article and your description will mean nothing to them. They will need some sort of quantitative measurements for your description to have any meaningful, practical use.

190. Diagram Your Terminology: One of my other big pet peeves is when a professional paleontologist writes up a description of a new species or genus and they use obscure terminology to do it. Sometimes they simply invent their own terminology to describe odd little bumps, ridges and structures they believe are incredibly important in defining the specimens' taxonomic position. The trouble isn't that they do this per-se, but that they often don't clearly define what it is they are talking about. We get it… you're smart. If you would really like the rest of us dummies to understand what it is you are trying to convey… draw us a quick picture or take a photograph and label it (well). It's not hard. What is the point of writing a scientific paper if only

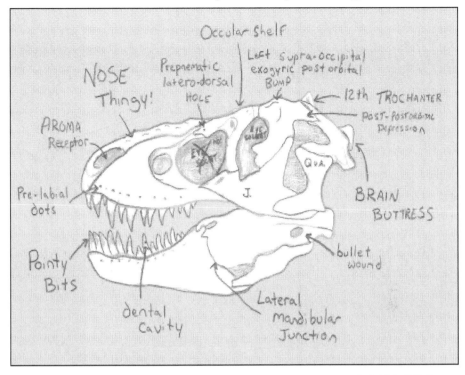

#189/190- FOR SOME STRANGE REASON, FEW PEOPLE COULD UNDERSTAND DR. EGGHEAD'S SCIENTIFIC DESCRIPTIONS. AS A RESULT, HE RESORTED TO DIAGRAMMING THEM, BUT TO HIS DISMAY, IT DIDN'T HELP MUCH.

5% of the people reading it understand it? Do not assume that every one of your readers will know the obscure structures you are describing.

191. Organize Your Data into Easy to Read Tables: One of the best ways to show comparative data is to place that data into an easy to read table or graph. Place measurements, ratios, descriptions or structures next to one another in a table. Try to make sure you explain where the measurements came from and/or

how they were taken somewhere in the text below. Try to make descriptions clear and concise, but detailed enough to be worthwhile. All scientific descriptions of new vertebrate taxa should have morphological measurements of key elements placed in tables so other researchers, who do not have the opportunity to view the specimen directly can compare your specimen to their own. When a reader can see your data, laid out, side by side, the differences or similarities become crystal clear. Sometimes a single table can do more for your arguments than ten pages of complex writing.

192. Make It Interesting: Paleontology is a wonderful, exciting, amazing, uplifting field. It opens your mind and expands the boundaries of time and space. Why is it that certain authors choose to write/speak in boring, monotone, overly clinical, high brow styles? If your topic is so damn interesting that others need to know about it, and you need to write about it, give it an interesting title. Give it a personal tone. Give it some life. You want people to read your paper and say, "Yes, I agree with that person's arguments and I agree with their conclusions." You want them to say "I understood exactly what he/she was talking about." You want them to say "I learned more from that paper than the previous ten that put me to sleep". Those should be your goals. There is nothing wrong with writing papers that are both educational AND entertaining. They can be both professional AND interesting. These are not mutually exclusive concepts. Make it interesting (even if the topic isn't so interesting). Find a way to grip your readers and pull them into the topic at hand. They will learn more and you will have fun doing it.

193. Use Lots of Pretty Pictures: People learn in very different ways. Some people learn better with verbal communication. Others learn by doing, with hands-

on projects or experiments. Many are visual learners. Your papers should, assuming the editors allow it, include multiple photographs, graphs, charts and other images to support your data. It's often said that a picture is worth a thousand words. To a visual learner, this is so true.

194. No Plagiarism: Plagiarism occurs when a person knowingly copies, word for word, text and reprints it without referencing the original author of that text. They are essentially stealing the work of others and claiming it as their own. This is a terrible thing. If you had done it in graduate school, and got caught, it was grounds for expulsion. What do you think the penalties are after graduate school?

195. Give Credit Where Credit Is Due: We are not lone wolves hunting in a desolate plain. From the moment we were born we have had help along the way (at least most of us have). There may have been enthusiasts or ranchers who found the specimens. There were field personnel that collected the specimens and their contextual data under a blazing sun. There were laboratory preparators that cleaned them and lovingly pieced them back together again. There were colleagues and reviewers, friends and enemies who studied the specimens and your paper and told you where you were right and where you were wrong. There were editors that pointed out all of the little grammatical and usage errors that you failed to catch. There were mentors and idols that inspired you. There were lots of family, friends, relatives, strangers and peers that helped the paper to become complete. Be sure to thank them for it. It doesn't take much. I'm not saying you need to turn your acknowledgements section into a 30-minute long *Oscar Award* ceremony speech. I'm simply saying that you

should give a short bit of thanks to everyone involved that helped get the paper to the finish line. Otherwise, you are implying that you did it all by yourself.

196. Work With Editors and Reviewers Not Against Them: So,

you have done your research, conducted your experiments, consulted with your peers, studied the opposing views and finally have written you scientific masterpiece. Now, all that's left to do is take in the accolades and fend off those who disagree right? Wrong. Your paper will need to be reviewed by many other people, before you will ever see it in print. Others will go through it with a fine-toothed comb, looking for errors in methodology, grammar, language, flawed logic, etc. etc. etc. Your work will first be reviewed by one or two colleagues (usually anonymously to ensure honesty), who will point out the papers flaws and make suggestions on how to correct them. Upon seeing such suggestions, it is often easier to throw a temper tantrum and throw the entire paper in the trash rather than take heed of their advice. Sometimes even the best of advice and most constructive of critiques can come off sounding negative or hateful. Remember though, their review was done to improve your paper. Their intent is not to destroy your self-confidence or insult your paleo-worthiness, it is to help you produce your best, honest work. Swallow your pride and follow their advice. You will find that you and your paper will be better for it.

197. Copyright and Trademark Everything: Modern day copyright laws have

freed the scientist from a lot of, once required, legal mumbo jumbo. Essentially, anything substantive an individual produces (text, artwork, music, ideas, concepts, etc.) is now considered copyrighted immediately upon its creation. You can make it more legal by contacting the U.S. copyright office and filling out the appropriate forms and paying a small registration fee. See www.copyright.gov for more information and copyright basics. Most professional papers submitted to journals are going to be copy written by the publishers of the journal. For other things like artwork, molds and casts, books, and the like, make sure your rights as the original creator are protected.

198. Investigate All Avenues of Publishing:
There are many places where one can publish research papers or trade articles on paleontology. These range from professional journals to popular science magazines, books, internet websites and forums. Your ideas need to be available in order for them to be tested. If you find one avenue is not open to your work explore another and another until you find the right fit. Sometimes self publication is the final straw, but hey, if Cope frequently did it back in the day, who can argue.

199. Trust No One:
I hate to pessimistic, but in this day and age, sometimes it's best to keep your ideas to yourself until you can refine them and complete them. Publication theft is a terrible breach of intellectual property rights. It unfortunately does occur from time to time. There are numerous historical examples of one paleontologist stealing an idea from a colleague and then publishing on it before the other even knows what happened. In paleontology, priority takes precedence. In other words, he who publishes first gets credit for the idea and he who publishes last is often forgotten. Be very careful with whom you consult for advice and reviews.

200. P.T. Barnum Got a Few Things Right!

Phineaus Taylor Barnum was an unapologetic, brazen showman and entertainer of the 19th century. He is best known for his shameless self promotion and for making his entertainment venues far larger than life. He created circuses, side shows, trade shows, an aquarium, wax museums and an oddity museum in New York City complete with hot air balloon rides and curiosities collected from around the world (including many real fossils). Barnum's creations were not always legitimate or even accurate (many were downright hoaxes), but his shameless self promotion always attracted big crowds, lots of media attention and for Barnum, financial and personal success. It would be easy to paint Barnum as an egotistical blowhard who duped people out of their money to see his bizarre attractions, but that wouldn't be the whole story and not the point of rule #200. You see, like Barnum, you need to be your biggest promoter. If you want to achieve great success in this field you need to promote your ideas and your skills with P.T Barnum style and flair. Don't get me wrong, I'm not saying to completely emulate the great showman and all of his more dubious tactics, but you shouldn't be hesitant when marketing yourself or your ideas. The most well known paleontologists do this all the time. That's part of the reason they are well known.

201. Never Alert the Media Until You Have Researched Everything Thoroughly:

Rule #201 is essential if you want to maintain your reputation. Too often, a paleontologist makes a seemingly big discovery and immediately runs to the media with a press release. Often, after the press has already run the story, and your peers are calling you up with kudos and questions, you discover that your initial interpretations were incorrect or perhaps, not quite as earth-shattering as you once thought. When that occurs, you make yourself look bad and you spread errors that never should have been started in the first place. So, make sure you have thoroughly analyzed your specimen or your project through and through before alerting the media. There is always time to go to the press. Make sure you are right before you sound the alarms and pop open the champaign bottles.

202. Conduct All Interviews Via E-mail or Written Correspondence:

When dealing with non-paleontologists in the media, I strongly recommend that you do your Q and A's through written correspondence. Newspaper reporters and magazine journalists are looking for the next big story (they follow Barnum's rule #200) and in their eagerness, ignorance and time constraints, can completely mess up an interview. They may not understand exactly what it was you where trying to say or they needed to spruce up the story to get it interesting enough to go to press. Either way, phone interviews often lead to embarrassing quotes and misleading, inaccurate stories. Over the last several years I have been burned by writers that just didn't take the time to make sure what they were writing was correct. So, as a result, I no longer do phone interviews. I insist that all correspondence be done over e-mail. It's much harder for them to misquote you or misunderstand a position if it's written down, right in front of them, as they are writing their piece.

203. Expect to be Misquoted: No matter

how careful you are with interviews, expect to be misquoted. Just accept it. If you follow Rule #202, you can alleviate some of the risk, but not all of it. Sometimes, it just happens. Reporters tend to get a few things wrong; an innocent error here or there, a misunderstanding, a phrase re-written to make your statement sound more exciting, that unfortunately makes it scientifically inaccurate... any number of things. In some cases, the reporter has a hidden agenda that they are trying to show in the piece. When this maliciousness occurs, they write only what fits their agenda and not the 99% of other useful information you gave them. A recent interview I did, with a very well-known science magazine, did just that. They wanted to make me and the commercial paleontological organization I was working with look bad because they wanted to discredit commercial paleontology in general. The interview went well, I thought, despite some of the more leading questions. Overall, though, I thought I argued our case pretty well and when I hung up the phone I was confident it would be a fair article. Weeks later, when the story came out, I was shocked by what I had read. The headline for this "news" story read; "New Journal Supports Black Market Trade in Fossils"! My jaw dropped to the floor! The article left out most of what I had said, skewed the quotes to fit the reporter's agenda and showed only what the reporter wanted to show. It was filled with blatant inaccuracies and even downright lies (no, in case you are wondering I do not support the black market trade in fossils). It's not right, in fact, it's downright evil, but it happens and happens all too often. Prepare yourself for it. If necessary, research your budding, young reporter to detect if they have any hidden agenda and demand that they provide a rough draft of the piece before you will allow your quotes to be used. If the reporter is interested in accurate reporting they shouldn't have a problem with that. If they refuse, it might be best to turn down the interview.

If you are doing an interview for a television show, expect the editors to cut your best lines and play the dumbest thing you said over and over and over again. When that happens, it's probably better to shrug it off and try to forget about it, than waste time in a futile protest. Move on and learn for the next interview.

204. Don't Take Yourself Too Seriously/ Leave Your Ego At The Door:

Yes, in order to climb the ladder of paleontological success, you will need to be a shameless self promoter (Rule #200). BUT... that doesn't mean you should be an arrogant, pompous ass either! Be respectful of your peers and respectful of yourself. If you take yourself too seriously and become insulting, condescending or ego-maniacal, expect others to try and cut you down. The moment you make a mistake (and you will... we all do), you will be forced to eat a healthy portion of humble pie.

205. No Two Paleontologists Will Agree on Any One Thing:

One of the beautiful things about paleontology is that its damn near impossible to find two experts who agree on any one thing. Present a strange, mystery bone you are trying to identify and you will have twenty different ideas from twenty different experts. Trying to piece together the mysteries of ancient life from busted up, crumbling, 180 million year old bones is difficult. Expect opinions to vary considerably.

206. Expect To Argue Frequently:

As I said before in previous rules, paleontologists generally love to argue. There are a few who are more civilized and prefer to quietly conduct their work out of the limelight in a more dignified manner, but many simply love to argue.

#207- Tommy DeRaptor was known to be just a little unstable with criticism."

Going to a professional paleontology meeting and listening to a controversial talk is like attending a *World Wrestling Foundation*, steel cage match. It sometimes gets ugly, but it's always entertaining. If not morbidly so. I recall an SVP meeting in Chicago in 1997, where opponents of feathered dinosaurs and supporters of feathered dinosaurs practically yelled at each other from the stage. If you intend to climb the ladder of paleo, be prepared to occasionally enter the cage match.

207. Don't Take Criticisms Personally- As the Mafia Says... "It's Just Bizness!" Constructive criticism is a great thing. Sure, it might sting your ego a little bit to admit your research paper had a flaw or two, but it is essential if you are to progress as a writer and a scientist. You cannot take these criticisms personally. In most cases, the criticisms are not meant to be personal attacks, just disagreements with your work. Instead of closing your ears and getting defensive, listen to the disagreements and learn from them. Then come back swinging in the next paper. If you are too shy or timid, or criticism tends to ruin your day for years at a time, I would suggest finding a nice comfy spot in your local museum basement and not publishing too much in this field.

208. Don't Let the Bastards Grind You Down: Rule #207 and #208 go hand in hand. I've seen many bright young paleo students get frustrated, disheartened and absolutely crushed by criticism (some of it was legitimate and some over the top). I know of one who recently decided to change professions because he was so sick of the petty bickering and callous criticism. That was a shame, for the young man had incredible skill

and potential. If you are going to survive, you must have or be able to grow a thick skin. You can not let your opponents defeat you. You will survive, if you follow rule #204 and not take yourself too seriously (be humble) and make it your mission that despite the odds and the consequences (or the jibes) to never give up.

209. Competition is a Good Thing: A

lot of people are afraid of competition. They like a nice, safe, protected environment, where they don't have to compete for dig sites, research projects, the best, most talented staff (field crews, prep techs) or the most esteemed jobs and duties. I disagree entirely. Competition makes us stronger. It helps us to understand our strengths and our weaknesses. If forces us to do a better job than we thought we were capable. It picks us up and keeps us moving when it would be far easier to quit or take the day off. Competition is wonderful. In paleontology, competition during the great bone wars led to the discovery of hundreds of new specimens and dozens of new species. Competition encourages the testing and re-testing of old ideas, making them stronger. Competition produces volumes of research instead of one or two papers on a subject trickling out per year. Encourage competition in and around your workplace and your workplace will be better for it.

210. Every Decade or so, Expect a Paradigm Shift: Paleontology is a very dynamic

field that is in a constant state of revision. It is a relatively young science where new discoveries are frequently made. Shifts of the dogmatic paradigm occur with relative consistency. Imagine, that just forty years ago we thought of dinosaurs as lazy, cold-blooded behemoths incapable of holding their heads above water, doomed to extinction. A mere thirty years ago, the idea that asteroids occasionally struck our planet with catastrophic, devastating force (including one that helped wipe out the dinosaurs) would have been blasphemy. Just twenty years ago, it was extremely controversial to suggest that some dinosaurs behaved in complex, social groups, possibly raising and protecting their young. A short ten years ago, it would ignite a firestorm of controversy to suggest that birds had evolved from dinosaurs and that we might find evidence of feathered dinosaurs one day. A mere five years ago, who would have thought that it might be possible to recover 65 million year old protein molecules and possibly even dino DNA? Certainly not I. This field changes faster than you can blink your eyes. Be ready to consider and possibly accept evidence for a few things that your professors would have vehemently scoffed at.

211. Think Long Term not Short Term- Think Deep Time Not Shallow: Paleontologists need to view the world in deep time. It's difficult for introductory students to begin to grasp this, but eventually it sinks in.

212. Before You Analyze Their Methods and Data Understand Their Motives and Assumptions:

Even the best scientific papers are founded on certain key assumptions. These are the premises from which the author(s) have built their arguments, determined their methodology of choice and approached their conclusions. Every scientist makes a few assumptions when doing research and writing papers. Sometimes they will assume the credibility and "truth" of a particular theory and indirectly use it to bolster their own paper. Perhaps they assume the credibility of a particular method and assume you, the reader, will also accept the method without question or even the slightest thought. Sometimes, an author will assume a particular taxonomic classification is valid or assume the work of someone before them was valid. The trouble is, the more assumptions a paper relies upon, the less strong the arguments and the conclusions. For example, you start reading a research paper that claims dinosaurs were slowly going extinct long before the K-T boundary. The paper is well written and the data seems logical, almost irrefutable. Then, you begin to look into the assumptions that the author makes in getting to the papers conclusion and you become less impressed. You discover that the author assumed that the radiometric dates used and cited were accurate even though they were reported 20 years ago with old technology. You discover that the author assumed a faunal diversity count of only 12 genera- this was based solely upon well known genera represented by complete skeletons only... not teeth, not indeterminate isolated bones, not unresolved taxa, etc. You also come to discover that the author chose to ignore and not even mention six other species which he considered invalid and assumed you wouldn't know about them. The author also assumed the invalidity of a dozen or more specimens which were owned privately and immediately discounted as a result. The paper assumes that the sample size is large enough to represent an accurate population cross section. The paper assumes that there is not collecting bias or preservational bias in the sample set. If the author is incorrect in assuming the dates were accurate, or the biodiversity count was accurate, or the sample size was sufficient or the questionable taxa were indeed valid, or their was preservational or collecting bias problems, then the whole paper can be thrown into question. It doesn't mean that the author's conclusions are wrong per se, but it does mean the evidence is far less than the author would like you to believe. You must understand what the author assumes to be true before you judge the validity of the author's conclusions.

Another thing you need to look for is an author's motivation. I hate to even bring this up as science should be above such petty stuff as personal motivation, but as humans we are all guilty of underlying motivations. You need to ask yourself, has this author made a career of defending this position? Have they any personal or financial gain from the paper or specific conclusions in the paper? Who provided the funding for the papers research? Who else might be connected to the paper that the author might have been influenced by? Who were the author's mentors or instructors and did they also argue these same points in the past? Does the paper support a particular political philosophy that may have influenced its conclusions? There are dozens of things which go to motivation. Why would "X" person, come to "Y" conclusion in the face of equally plausible evidence? Motivation does not in any way invalidate good, solid research, but when there seems to be a hidden agenda or a not-quite-so-legitimate motivation for a paper's conclusion, you need to analyze the data even more carefully then normal. Motivation is a difficult thing to interpret (and prove for that matter), but something that should be considered.

213. Be Careful Whom You Criticize and How You Approach it; It's A Small Paleo-World: A few years ago, I had

the pleasure of attending a paleontology conference in Illinois. I was sitting in the back listening to the speakers and taking notes. One of the speakers that day was a good friend of mine, and he took the stage that afternoon to deliver his presentation. While I was listening to him speak, a young couple behind me, just within earshot, began gossiping about my friend on stage, leveling all manner of nasty, personal attacks amid whispers and giggles. The two hecklers were not paying much attention to the actual presentation or discussing his points pro or con, they were simply tearing down the person because of who he was and who, they perceived him to be. They had no apparent concern for who might have been nearby overhearing their baseless and vicious attacks. They certainly did not know that the person who was sitting right in front of them was listening in and just happened to be good friends with the person they were tearing down. I could have turned to the two and asked them to please be quiet or defended my friend's honor with some sharp words of my own, but what good would that have done? You can't cure stupidity. So, instead, I quietly listened, fumed, and made a mental note of the hecklers' faces and later, by asking around, discovered their names. Now, you can bet, that should I encounter them in the future, I will more than likely treat the pair with disdain and distrust. If I am ever in a position to help them, you can also count that no help will come.

Rule #213 reminds us that the world of paleontology is rather small and that most people who are involved in it know each other pretty well. We may not always like each other, but as professionals we should at least give each other a certain degree of respect and civility. You need to be very careful about whom you criticize, how you criticize them and whom you criticize to. Constructive criticism is good and expected. In many cases it is demanded. Arguing, debating and wrestling over ideas and concepts is perfectly acceptable. Ad homonym attacks and silly gossip are not. Remember... that lab-geek that you and your friends are making fun of, just might be in charge of a museum that you might want to work at ten years from now. The field gal who you treat terribly on the dig site could wind up on a review board that determines whether or not your book gets published. That guy you made fun of in college, might wind up deciding whether or not to grant you a federal access permit or write you that next check for your research. Keep those things in mind the next time you feel like playing the class clown.

214. Do Not Be Blinded By "The Call To Authority": One of the most dangerous logical fallacies prevalent in paleontology today (and many other branches of science and culture) is known as "The Call to Authority". This fallacy of logic states that because a person is in a position of authority (higher degree, higher position) then whatever that person says must be true. BBZZZZ! Wrong answer. It doesn't matter who you are or how many degrees you hold. If you say something that is inaccurate, it's inaccurate. A lot of people in positions of power use this fallacy to enforce their directives (because they know best) and drive their agendas. Don't be fooled by this. Just because your mentor says something, doesn't make it true. Just because Dr. so and so, with 40 years in the field argues for X, doesn't mean that X is true or the only possible hypothesis. A good scientist remains skeptical.

215. Sorry, But There is a Little "Faith" in Science: A few years ago, I was having a late-night, beer-induced, philosophical discussion with some close paleo-friends on the topic of "faith" in science. My friends, were in a rare militant mood and were rambling on and on about the irrationality of organized religion. Again, this is all well and good; we had our philosopher's robes on at the time, (see rule #11) but their arrogance struck a nerve in me and I decided to play a little devil's advocate. They claimed that faith had no role in science and that they did not rely upon it, in any way, in their work. I countered with a few questions that

they were hard pressed to answer. One of those questions went something like this: "Have you ever personally conducted a radiometric date? No? You? No? Neither have I. But we assume that the workers before us, who have sampled the rock and processed those samples did it accurately. We assume there was no contamination. We assume they did the math correctly. We assume that those layers have been tested and re-tested and that the numbers were reported accurately. We have FAITH in their results and their methods". We could conduct the tests ourselves and personally verify the results, but how many of us actually have the time to do this? Very few. Therefore, we must have faith in our fellow scientists that they are accurately and honestly reporting and recording their data. If we do not have a least a modicum of faith, our inability to accept ANYTHING would cripple us. So, faith does exist in science at least to a small degree. We stand upon the shoulders of giants. We must remain skeptical about their work, but if we have no faith whatsoever in their methods and observations we can never move forward.

216. Beware of Science as the New Religion: Okay, this one is sure to get me into trouble. Just bear with me a bit, before you burn me at the stake. Whereas there is some faith in science, and it is in many ways a necessary evil, some paleontologists take this to extremes. Over the course of my many years working in and around paleontology, I have encountered many, so-called professional scientists, who treat their politics, theories and hypotheses like fundamentalist, evangelizing bible thumpers. They have somehow strayed from the true nature of the scientific method and crossed into the dangerous, militant world of the advocate. For them, it is no longer just about arguing for or against a position, it is about declaring, beyond any reasonable doubt, the "truth" of their position. These proponents

bitterly and fervently attack any opposition of their logic as if they were on a crusade against the infidels. These proponents even have their share of cult-groupies who follow them about looking to help "prove" their master's ideas. For them, it is no longer "science" it has become "belief". Many, particularly the militant atheists of our creed, argue that science can and should know everything. This is, I fear, a dangerous position to hold. As scientists we must always have an open mind. New data can throw our ideas out the window faster than you can say "Oops!". If we fail to accept new data, if we blindly accept things on faith, then we are no longer practicing science and have strayed into the quicksand world of belief.

217. Life is Always More Complicated than Science Admits:

Paleontology loves to try and pigeon hole organisms and natural processes into nice, easily defined boxes. Our taxonomists run their cladistics programs and then choose one, most parsimonious tree out of a dozen potentials and call that the correct one. Sometimes, convergent evolution throws us curveballs making two completely unrelated organisms appear to be closely related. Other times rapid adaptation or speciation makes two closely related organisms appear to be distant. Sometimes sexual dimorphism is extreme in the natural world, leading scientists 70 million years later to conclude there were two separate species instead of two sexually distinct morphotypes. Climate changes in ancient systems may have been effected by solitary causes, but more than likely were caused by any number of complex factors. Mutation rates and survivability may have been uniform or may have varied widely depending on genetics, solar activity, temperature, environment, chemistry and any number of factors. Extinction events may have been single cause or multi-causal, random and chaotic or ordered and cyclic.

Life is complex. It is not easy to classify and pigeonhole. Do not assume that your breakthrough scientific paper is taking into account everything in those complex ancient ecosystems. See also Rule #183.

218. Life Finds a Way: Evolution and adaptation has produced a wide array of truly bizarre creatures. Dinosaurs in particular, have an ungainly assortment of frills, horns, armor, clubs, crests, plates and spikes. Some of the organisms, when you look at them, seem to be so incredible that it's hard to imagine the organism being terribly successful. For most of paleontology's history, scientists viewed these odd beasts as evolutionary dead-ends. Some even argued that the dinosaurs went extinct due to racial senility; i.e. they just got too weird for their own good! I recall a few years back, that some paleontologists were studying the long necks of sauropods and trying to prove it was mathematically impossible for them to hold their heads higher than a few feet in the air, or, they claimed, the beasts would faint and collapse. Other scientists made the bold statement that *Tyrannosaurus rex* was an obligate scavenger, living solely on dead carcasses, predominately because they believed the tiny arms of a rex would be too small to use in hunting and its legs were too fat for fast running. These ideas, and many more, which I have not mentioned, are all ridiculous and easily dismissed when applying Rule #218 correctly.

Life finds a way. We do not know precisely what many of these unusual adaptations were for or how the organism benefited by them (or were not affected by them at all, or were detrimental). But, we do know that they had them and for some "bizarre" reason they continued to eat, mate, breed, raise their young, migrate, adapt to their environment, ward off predators, catch prey and somehow survive through thousands of generations and millions of years. Those who do not grasp Rule #218 simply cannot accept this and wonder, "How is this possible?"

We don't know exactly how they did it, but it worked for them. They most certainly weren't running around passing out every five minutes! They weren't tripping over their two left feet and breaking their necks! (How many ostriches running 60 miles per hour without "arms" trip and break their necks?) Why have stereoscopic vision, excellent smell, strong powerful legs and a mouth full of 12 inch-long, banana-shaped daggers, if all you need to do is feed on dead carcasses? You don't need to be fast to be a good hunter... you just need to be faster than what you eat! Why have long necks if you can't raise them to reach the tastiest leaves in the trees? Most successful animal groups' have some "bizarre" adaptations and these generally imply some previous or current function. That function somehow made the animal successful at some particular point in time. Certainly long enough for it to become widespread within the population/group/species and so equally bizarre. Hypotheses that point to those "bizarre" adaptations as being the cause of their demise, without further evidence always need to be critically examined.

219. Form Follows Function: Each bone, muscle scar, individual structure, bump, ridge, crest and bizarre adaptation implies a specific function or use. The morphology of that adaptation can help us through simple mathematical formulas determine the capabilities of that use. This study is known as functional morphology and it is widely used in paleontological studies. Through functional morphology studies we can determine the probable speed of an organism, the flexibility of its digits, its capacity for flight or skill in swimming. We can determine its limb strength and its biting force. We can determine its leaping ability or climbing ability. We can determine what the animal may have eaten or where it

Darn it! Hey everyone! Hold Up! Stop the Hunt! Stop! George Tripped again!

#219- EVOLUTION WEEDED OUT ALL THE CLUMSY DINOSAURS... EXCEPT FOR T. REX... WHICH WAS WELL KNOWN TO HAVE TWO LEFT FEET.

may have preferred to live based upon its morphology. So, whereas Rule #220 will point out the fallibilities of using morphology, rule #219 shows us that morphology is a very useful tool for analyzing many aspects of a fossil organism's abilities and possible behaviors.

220. Paleontology's Morphological Achilles Heel:

In most cases, all that is left of the animals we study are their hard parts; the bones and teeth. The morphology and position of those hard parts can help tell us how they are related, but morphology alone is never going to give us an absolute answer. Unfortunately, without a time machine, it is in most cases, our ONLY method. Biologists studying extant organisms have the opportunity to go into an extant animal's native habitat and study how it interacts, how it breeds and which individuals it successfully mates with. The traditional definition of an extant species is: "a group of organisms that can successfully mate with one another and produce viable offspring". Biologists, for their animals of choice, can effectively tell this by direct observation (they know the momma, they know the poppa, they see the babies). Paleontologists, for their animals of their choice-fossils, cannot. We can't see if Daspletosaurus could successfully mate with Albertosaurus. We can't know whether or not a mastodon could mate with a mammoth, or whether a cave bear could successfully mate with a short-faced bear. For paleontologists, determining which animal belongs to which species is limited to morphology. If we look at today's mid-sized, pack hunting, loveable little household carnivore, the dog, *Canis lupus familiaris*, we can easily see the problem with using morphology alone to determine species. The skeleton of a Chihuahua is dramatically different than that of an English Bulldog or an Irish Setter. The morphology and shape of the bones are very different and yet, each of the over 800 breeds of dog, are in fact, of the exact same species. They can all interbreed successfully. If we were to see some of the various morphologies of *Canis* preserved in the fossil record, we might be inclined to call some of them distinct genera. Beware of the limitations of morphological classification. I'm not saying not to use it. I'm merely pointing out its limitations.

221. Individual Variation can be Great: Shaquille O'Neil or Danny DeVito?

The odds are that the specimen you are currently working on is going to be an average, adult individual. It is, more than likely, going to be a representative of the normal sized, average morphology and have all the typical characters of the entire species. On rare occasions though, you may have found an individual that doesn't fit the average and falls more towards the extremes in size or morphology. Perhaps you found a giant or a pygmy? Perhaps you were lucky enough to discover the rare specimen with a genetic abnormality or some other bizarre morphological variation. It you have a large sample set available to study, you may be able to discover that your specimen isn't just average. If you do not have a large sample set to study (which is normally the case), you may be fooled into calling it a new genus or species. This description of your now "type" specimen, assumes that the type represents the average- the norm- not the extreme. That description, in this scenario will cause confusion in the field for years to come, pending further discoveries. The point is, be aware that individuals can and often do, vary greatly.

222. Evolutionary Progress or Chaos?

In the late 19th century, a great many scientists took Darwin's theory of evolution to mean that organisms evolved in a positive direction. That is to say, those organisms, over time and dozens of generations, "improved" their species characteristics. They became more advanced- better, smarter, stronger, larger, more complex, etc. over time. They believed that this evolution would be progressive and linear. A modern understanding of evolution understands that this is not usually the case. Evolution does not work in a linear direction where future generations are without question "superior" or more "advanced" than ancestral generations. There is no biological progress. What is beneficial to an organism at one particular point in time could be detrimental at another future point in time or in a different environment. Time, mutation, competition and environmental change are the four factors that determine whether an adaptation is beneficial or detrimental. As environments change, the usefulness of an adaptation may become obsolete or in fact become damaging. During the ice age, the thick hairy fur on a mammoth's backside was a wonderful adaptation that enabled the animal to be warm and quite successful. It enabled the animal to live, reproduce and pass its genes forward. Those genes however, and that thick fur, started becoming detrimental to the animal as the environment began to warm up.

223. We Do Not Know All There is To Know and Never Will: Paleontology can teach us amazing things about the world, its life forms and our shared history. It does, however, have its limitations. There are many things about the history of this planet whose record has been permanently erased through the natural processes of erosion and decay. Large gaps of time have been wiped out. Entire species and groups of animals have vanished without a trace. There will always be gaps in our knowledge. Those gaps do not make the field less important than other sciences, but should give the scientist caution when overstating what we can know.

224. 99% of Life Never Fossilized: The fossil record is terribly incomplete. The majority of this planet's creatures, lived, reproduced and died without their lives ever being recorded in the rock record. Fossilization is a very rare event. In order to become

fossilized, organisms need to be buried quickly in the right types of environment. If they are not buried quickly, their soft parts rot away and their hard parts slowly turn to dust. Ancient life was probably just as diverse as modern life, we simply will be missing some of that diversity. Organisms that lived in past environments where erosion dominated and deposition was slim (like high mountain habitats, canyons, rugged terrain, forest land, high plains, etc.) had little chance of being preserved. Whenever you are doing any research project in the field of paleo remember Rule #224 and the inherent problems of the rock record.

225. Are You a Lumper or A Splitter? Better Figure That Out Soon: In paleontology, particularly the taxonomic part thereof, there are two main types of scientists; those informally known as "lumpers" and those informally know as "splitters". A lumper is someone who, when studying a new specimen, tries very hard to place the specimen into previously established taxa. Any variations in morphology or structure they tend to account for as individual variation, sexual dimorphism or ontogeny (age variation). As a result, lumpers are very reluctant to name new species or genera, and when they do, it is usually for very good reasons. When they don't, their analysis needs to be carefully analyzed to determine how accurate their interpretation is. Splitters, on the other hand, when studying a new specimen, have the tendency to see minor variations in morphology or structure as vitally important, generic level or species level changes. Splitters tend to put less stress on individual variation, sexual dimorphism and ontogeny and love to name new species and genera. As a result, when they name a new genus their interpretations need to be extra carefully assessed, to determine if their analysis is the most likely one or if they are just naming something new for the sake of naming something new.

#226- ALL OTHER PREVIOUS ATTEMPTS AT COUNTING DINOSAURS HAVE FAILED MISERABLY!

When a scientist who tends toward being a splitter chalks a morphological change to individual variation, sexual dimorphism or ontogeny, their analysis is probably the correct one (of course like anything in science it still needs to be carefully scrutinized!).

As you begin to conduct your own research and begin to ask such taxonomic questions for yourself, you will find that you will tend toward one of those extremes. Sometimes your tendency will change as your knowledge base increases. Self realization will help you to understand when your tendency to be a lumper or splitter is negatively affecting your research.

226. You Will Never Understand Extinction by just Counting Specimens: As we have seen in rule #224, the fossil record is terribly incomplete and in rule #186, absence of evidence is not evidence of absence. When analyzing any extinction event, simply counting specimens will not tell you, by that data itself, much about an extinction. Can we really say, without any doubt, that dinosaurs are declining in the last 20 million years of the Late Cretaceous just because in the Judithian we have 27 valid genera and in the Maastrichtian we only have 22 genera (don't quote those numbers)? We could say it, but it would not necessarily be true. Complete specimens, preserved well enough for proper identification are rare. Genera counts are subjective and vary greatly from author to author. New genera are found all the time. Isolated bones and teeth suggest that there are many new species/genera yet to be discovered. Entire families of vertebrates may be so rare in the fossil record that finding them is next to impossible. There could be preservational bias or even sampling bias which distorts the numbers. Even if we had an accurate representation of the

biodiversity of that time period, can this alone be used to judge the cause or rate of extinction? Probably not, since one would expect biodiversity levels to fluctuate naturally. Rule #226 is inspired by Dr. Dale Russell who said a similar thing way back in 1994. Few people listened then either.

227. The Present May be The Key to the Past, But the Past is the Key to The Future: Charles Lyell in his 1830 work, *Principles of Geology* developed a series of ideas now known as the Theory of Uniformitarianism. This theory is often stated as; *"the present is the key to the past"*. In other words, the same natural processes that are operating today, at slow and steady rates, over millions of years, can be used to explain all of the geology we see in the rock record. Therefore, if we wish to determine whether or not a particular section of rock represents a lake environment, we go to a modern lake and observe the sedimentation rates and look at core samples of its bottom sediments. If the modern lake sediments match the sedimentary rock from our suspected lake (in color, texture, mineral/chemical composition, etc.) then our sedimentary rock indeed was deposited in a lake. If we observe earthquakes, volcanoes, erosion, deposition, mountain building, subduction, etc. etc. etc. occurring today, then it likely occurred in the past. Again… "The present is the key to the past". We do not require any unknown and no longer operating processes to explain our geology. Despite many flaws in Lyell's original work, Uniformitarianism has pretty much stood the test of time and is used by geologists all over the world as an operational theory. The flip side of this is also true however… "The past is the key to the future". In other words, we can use the historical rock record (and the fossils included therein) to help show

us patterns and processes that occur over long periods of time. Once we understand how those processes work over long time periods, we can then predict how those same processes acting today will change the surface of our world in the future. If we want to understand the reasons behind normal extinctions and mass extinctions that may occur in the future, we need to understand how the process worked in the past. If we want to know when the next earthquake is going to strike we need to study the historical movement along the same fault zone to aid us in our prediction. "The past is the key to the future".

228. There is a Grain of Truth in Every Myth; The Trick is Finding That Grain: Myths of dragons, sea serpents, griffins, Cyclops and other monsters exist worldwide in nearly every culture. To our modern, scientifically trained eyes, we often discount such stories as the over-imaginative minds of our ancient ancestors. Many dismiss them and move on to seemingly more important things. Rule #228 has been included here, because I think it is important for the paleo-student to consider such myths for longer than just a brief second, for those myths just might contain some interesting information about our fossil friends. The ancients were not dumb; they simply had very different terminology and a very wise fear of the dark. Perhaps, the ancients encountered the bones of dinosaurs and in their limited knowledge described them as "dragons". They would see the skull of a mammoth, and think they were looking at the skull of the monster Cyclops (the hole in the skull for the trunk looks amazingly like a single orbit). A wonderful book on the subject of fossils in the ancient world, called *"The First Fossil Hunters"*, by Adrienne Mayor, outlines how these myths can lead us to actual fossil treasures. As students,

do not necessarily dismiss any myths out of hand. Ancient myths or modern ones. You never know, it might lead to the next big discovery.

229. Extinction, Like S--T, Happens!

Extinction is a fact of life. We hate to think about it, but one day even the great human race will disappear as the dodo. Sea levels will rise and fall. Mountains will crumble to the sea. Animals great and small will vanish forever. The average lifespan of vertebrate species on this planet has been estimated to be around five million years. That's not much time, a mere drop in the bucket so to speak. So, most of the species that were alive are now extinct and all of the animals that are alive today will, one day, be extinct. Sorry.

Today, we are told by numerous experts that we are in the middle of one of the worst ecological disasters and extinction events this planet has ever faced. Biologists, often with the aid of certain paleontologists, geologists and paleo-climatologists claim that we are loosing species at unprecedented rates. They have altered the traditional definition of species (splitting species by geographical isolation rather than the ability to mate and produce viable offspring) to raise the estimated number of worldwide species. As a result, they sometimes inflate their extinction numbers. Sometimes they will invent a new name for a species of insect (on little evidence), and then suddenly claim it has gone extinct, without bothering to check under every branch and leaf in the jungle to see if it is truly gone. When it comes to politics, do you really think they are looking that close? Some estimates I have seen suggest something like 2000 species a day are going extinct in the rain forest! I'm not exactly sure how that is possible, for at that rate, all life on the planet should have been extinguished by last Tuesday!

They blame human encroachment and the destruction of natural ecosystems. They look to our

#230- IN THE LAST DAYS, AS THE CLIMATE CHANGED AND THE WORLD BECAME COLD AND DARK, THE POPULATIONS WERE *ENCOURAGED* TO DRIVE SUV'S. UNFORTUNATELY, THE INFRASTRUCTURE JUST WASN'T THERE!

changing climates and have produced the hypothesis of global warming, a "crisis" that many have suggested will melt our polar ice caps and raise sea levels by 20-150 feet before the end of this century. Many of the radical types in their ranks suggest that thousands of species will go extinct, millions of people will die, world societies, cultures and economic systems will collapse and the survivors will fight for crumbs in the remnants of a *Mad Max*–like, barbarian world unless we all bow down to their will and follow their politics. For many, the sky is falling and human beings are to blame- end of debate!

While human beings ARE notoriously bad at taking care of the planet, this apocalyptic vision paints our future in a very bleak light making the average Joe and Jane contemplate whether it's even worth waking up in the morning. However, before you start building your fallout shelter and stocking canned goods it might be best to stop, take a deep breath and think for a moment. Extinction events have occurred in the past and they are most certainly going to occur in the future. The largest extinction event known, the Permian-Triassic, is thought to have eliminated over 90% of life (on the family level) on the planet. The second largest, the K-T event, 65 million years ago, took out an estimated 65%. Now, those were extinctions! Human beings had nothing to do with either of them.

I do NOT know whether global warming is indeed a long term change in our climate caused by human beings or if it is merely a natural cycle running its course just like all of the other swings in average temperature for the last 570 million years. It is certainly worthy of serious, responsible study- not scaremongering. With that said, however, I do know a few things that should be stated: 1) the planet Earth was, for most of its 4.65 BILLION year history, significantly warmer than it is today. Even with today's so-called high temperatures, we are still well below the norm. 2) The ice ages and our little temperate, inter-glacial periods (we are in one today according to many), are more of an aberration rather than the norm. 3) During the Late Cretaceous the entire inner portion of the North American continent was underwater in a shallow sea as sea levels were significantly higher than today. There were no ice caps (no, the dinosaurs were not driving *Hum-vees!*) 4) I know crocodiles once lived in South Dakota, palm trees grew in abundance in parts of Canada and dinosaurs were living comfortably from Alaska to Antarctica. 5) I do know, that had we begun this debate at the start of the "little ice age", a mere 150-300 years ago (est. 1650-1850), that we would be worried about global cooling and not global warming. (looking at the ice core data, I am actually more concerned about global cooling). 6) I do know that if you live in a city that is eight feet below sea level, you can pretty much bet that it WILL be under water, sometime in the near future, irregardless of how big you build the levees or whether we all go back to a Stone Age way of life. In fact, once our long cooling trend is over and climates return to the global pre-Pleistocene average, the sea levels will rise and any city at 0-150 feet above sea level will probably be in serious trouble. 7) I do know, that some species are going to go extinct no matter what we as caring human beings try to do. The panda, the spotted owl and the like probably don't have too much time left... sorry. 8) I do know that as we tamper and interfere with natural processes, we can produce unexpected negative consequences that we, may or may not, live to regret. 9) I do know that is not wise to build a city on swampland, unprotected coastline or an earthquake-prone, fault zone. Sooner or later you will get hit with a natural disaster. Duh! 10) I do know a single, large volcanic eruption, like those that have occurred frequently in the past, has the potential to belch out more CO_2, water vapor and greenhouse gases than the last 150 years of the industrial revolution combined. 11) I know a few other things, but I've rambled long enough. The point is simple... before you, as a young paleontologist, are recruited onto the chicken-little, tie-dye-colored, bandwagon, and begin trying to outlaw my SUV and encourage our government to send all of our tax dollars to the United Nations for re-distribution (so we can ALL live in third world countries), please take a few moments and study the Earth's climate and extinction patterns over DEEP time. Not the last 100 years. Not even the last 30,000 or 70,000 years. Start looking at the cycles in MILLIONS of years. Learn about extinctions and the processes that have caused them in the past. Understand the statistics, the methods and the agendas. Only then, will you know when, where and IF to build your high mountain, Armageddon proof bunker.

230. Dinosaurs Were Not Meant to Live in Basements: Many museums and

universities, all across the world, have basements full of fossil vertebrates. Many of them are still sitting in their original field jackets just waiting for a young, enthusiastic paleo-student to come along and show them the light of day. Some of the American ones have been sitting there since the late 1800s. Many contain new species and genera that have yet to be described. They don't need to be buried in a dark warehouse. They should not be ignored, lost and forgotten. They need to be studied by paleontologists and placed on display where kids and adults can learn about them and stare at them in wonder. They need to be properly curated in storage cabinets where scientists can measure them, study them, photograph them and publish on them. There are plenty of bones and plenty of eager people willing to work on them and learn from them. So why aren't we?

VI. Rules For Teaching Paleontology

Teaching is one of the primary components to a career in paleontology. It doesn't matter whether you wind up teaching in a university lecture hall or a second grade class. Teaching and communicating your ideas effectively are essential. If you are a museum docent, you will be teaching museum visitors. If you are a laboratory supervisor, you will be training staff. If you are a research scientist, you will be lecturing to your peers about your research. Unfortunately, most degree programs in the world of paleontology do not teach their students HOW to teach. The following bits of advice will not substitute for education coursework, but they may assist you when presenting your ideas to others.

231. Kids Love Paleontology- Use That:

Paleontology is a wonderful way to get kids involved in the sciences. Dinosaurs, flying reptiles, saber tooth cats and other monsters of Earth's past are an immediate draw to most children. I'm not sure whether it was their immense size or the fact that they once lived and are now extinct, but these once living monsters inspire kids to learn. Use them to your advantage in the classroom. Use pictures of dinosaurs to teach kids sorting and classification. Use dinosaur picture books at story time. Have them calculate the size of the animals. Have them calculate the speed of dinosaurs from sets of footprints. Ask them to tell whether or not a tooth came from a plant-eater or a meat-eater and why. Use dinosaurs

#232- Pssssst! Yo... Egghead? How do you spell "Thanatoconosis????"

to teach about geology and biology. Use dinosaurs to talk about extinction and evolution. You can use these wonderful animals to teach many principals of science. In fact there are entire books with suggestions on how this can be done. Unfortunately, in today's educational systems dinosaurs are merely a footnote. Children are taught the bare basics in most second grade classrooms but after that, dinosaurs as a group are usually abandoned. This should not be the case.

232. Teach to Your Audience: Interest in paleontology is not limited to a bunch of rabid second-graders. Whenever you are giving a lecture or presentation to various groups observe who they are, their ages, their skill levels and their level of interest. It will not help your students if you are teaching at a level far above their heads or well below their skills. Either would be intellectual suicide. Teach TO them not AT them. If you are teaching to younger children then use softer vocabulary and an excited tone. Repeat things often to make sure the younglings understand the concepts. If you are teaching to a diverse group of museum visitors, then give a general talk, but also mix in a few advanced things for those who can grasp them. Never, ever, talk down to an audience! Try to be on their level. Talking down to them is insulting and your students' minds will shut down and become unresponsive to your words. A good rule of thumb is to put yourself in their shoes and try to imagine what kind of things they already know. Then review bits of that to draw them in and then take them one or two steps beyond. Make sure they grasp the foundation of a concept before you move on. Never assume they have the foundation to begin with. No matter who your audience is, don't be surprised if most of them only capture and retain 10% of the material on the first pass. That's about average. Ten percent isn't great, but it's better than none.

233. The Importance of "Why": One of

the most essential things teachers often forget about when trying to communicate the finer points of ANY subject, is explaining the "WHY" part. Sure, you can teach them what kinds of animals lived in the past, how they lived, where they lived and the processes that acted upon them. But did you ever stop and explain WHY it was important for your students to know any of it? Most educators, unfortunately, do not. If they do decide to explain the "why", often it is done hastily and carelessly.

I remember sitting in a college calculus class, intently watching the professor madly scribbling equations on the chalkboard and rambling on in a dry tone about what it was he was doing (with his back turned to us and standing in front of his scribbles, of course). As the chalk dust was flying in all directions, I remember thinking that the whole exercise, for me, was pointless. I remember saying to myself, "I'm going to be digging in the dirt all day long, why do I need to learn calculus? This is such an incredible waste of my time!" At that point, the moment those terrible thoughts crept into my tiny, little head, my mind shut off to mathematics. I became a rock. I was incapable of learning anything the professor was rambling on about. Consequently, I never was able to grasp calculus and barely passed any subjects that had anything to do with it. To this day, the mere thought of calculus (and that particular professor for that matter) boils my blood. The professor failed to thoroughly explain the importance of calculus and how I might be able to use it in my chosen profession. One could make the grand case (and many have tried), that it is the student's responsibility to learn the material and discover the "Whys" for themselves. But this argument doesn't hold much water. It is the teacher's job to teach. That is what they are being paid to do. I would argue that if a teacher fails to explain why a student needs to know what it is they are trying to teach, then they have failed as an educator.

Why are YOU trying to teach them this subject? Why do they need to learn it? What benefits are there for them to learn it? Why is it an important subject? Why do they need to know it? We live in a fast-paced, knowledge-rich society. Each and every one of us is bombarded by facts, opinions and theories on an hourly basis. Our time is limited. Our cerebral storage capacity is finite. We are constantly, subconsciously picking and choosing which bits of information we are going to save and keep in the front of our minds. We cannot know everything. How can you expect your students to learn a subject if they have no idea WHY they should? They can't and if you don't explain it to them correctly, they won't. Take a great deal of time explaining to your students WHY they need to know what it is that you know.

234. Find the Child in You: Remember when you were a kid and you couldn't wait to get to a particular class (or learn about a particular class subject)? You would get so excited your belly felt like there were a million moths fluttering about inside. You went to the library and checked out a dozen books on the subject to prepare. You talked about it on the playground. You joked about it. You even lectured to your parents at the family dinner table. You couldn't wait for your teacher to get to that part of the curriculum. No? Hmmmm. Then perhaps you have forgotten what it is like to be a kid.

Children are excitable. If they are encouraged and cared for, if they understand the "why", they are enthusiastic learners. They are sponges for material- clay to be shaped and molded. Most children love to learn. Yes, I know there are a few bad apples- a few damaged goods out there, but those are the exception rather than the norm. Rule #234 emphasizes that as an educator you need to be just as eager about teaching as your students are about learning. More importantly, even if your students

#235- Mr. and Mrs. Theropod were pleased to see Jimmy finally taking an interest in school. They were just hoping for an interest that was more... well... "Significant".

are not eager to learn your material, then you have to be even more enthusiastic about teaching them. Approach your subject with child-like enthusiasm. You love your subject, otherwise, why do it? Express that love- that passion you have for the subject to your students. It doesn't matter if you think doing so is corny and un-cool. Some kids may laugh from time to time about your passionate delivery, but they will pay attention and they will learn. There are far too many educators out there who have lost that childhood enthusiasm and the quality of their teaching suffers because of it. You probably remember that professor who should have retired twenty years earlier don't you? You dreaded going to that class. When you did, you brought an alarm clock… just in case. Likewise, you also remember that wild-eyed, passionate professor who seemed to be on the verge of insanity for their love of their subject. You couldn't wait to get to that class each week just to see what kind of insanity would follow. If you find yourself teaching, in any format, be the wild-eyed one.

235. Education and Entertainment Go Hand in Hand: I can't tell you how many times I have heard from educators that it is not their job to entertain their students. This is nonsense! Yes it is your job to entertain your students. Your job is to teach and you can't do that successfully if you do not make the learning enjoyable. I'm not saying you should put on a comedy show or a juggling act when you are trying to convey the principles of evolution. What I am saying, is that your students will be able to pay more attention to a professor who makes the discovery of those principles interesting. You need to be fun. You need to be aggressive. Walk around the classroom and look each of your students directly in the eye. Talk to them- not at them. Make it humorous. Make it personal. Tell them an anecdotal story

that builds up the information. Use metaphor and analogy. Supply personal evidence. Use hands-on specimens that they can see and touch. Make your presentations entertaining and you will find that you will enjoy teaching more and your students will learn more information.

236. Students are Diverse. They Learn in Different Ways- Therefore, Teach in Different Ways: Every student

is different. Each one has different strengths and different weaknesses. As a teacher, you will need to understand, by observation and experimentation, which ways your students learn information. There are three main types of learners: visual learners, auditory learners and kinesthetic learners. Visual learners learn by observation. They pick up concepts by seeing them in physical examples. They love drawings, photographs, diagrams, maps and images. They love to read and need time to digest what they have read. Auditory learners learn by listening. They can focus their attention by homing in upon your voice. They will detect the slightest nuances in tone and pitch. They love to talk and discuss principles and hate to take down notes. You will need to give them plenty of opportunities to speak, participate in discussion groups and present their opinions. Kinesthetic learners, on the other hand, learn by doing. For them, concepts become clear when physically handling and manipulating objects related to the lesson. They love hands on projects. They love experimentation. They love demonstrations and practicing techniques. Whereas, a visual learner might understand a diagram better and an auditory learner might need to listen to an eloquent lecture, kinesthetic learners will need to take apart your laboratory equipment in order to understand how it works.

Every group you encounter will be composed of the three main types and possibly shades in-between. As an educator, you must be able to recognize the signs which indicate what type of learner each student is and use appropriate methods when directly working with that student. For large groups you should use elements of all three teaching methods.

237. Is This a Fossil? Is This a Fossil? Is This a Fossil? Even if you aren't an expert

(yet), expect people without a paleontological background to treat you as one. Everyone and their grandmother will bring you all manner of rocks, sticks, plastic saber tooth cats, etc., etc., etc., for you to identify with the inevitable question of "Is this a fossil?". Most people involved in paleontology will hear this question a few times per month if not more. Kids are particularly eager to please and will bring you all manner of things they have found in their backyard to impress you and try to get your attention. Many adults will as well. I once had an elderly gentleman bring in a metamorphic rock (schist) that he insisted was a fossilized dinosaur embryo. I tried to tell him that it was just a rock with some interesting banding that looked like a dinosaur embryo, but he wasn't convinced. Expect this question to come up frequently and try to respectfully and patiently explain what it is they are seeing. Do not dismiss potential fossils out of sight, because many times, what the person is bringing you actually IS a fossil. In the right neighborhoods, it may very well be an important fossil.

238. If You Do Not Know an Answer, Do Not Pretend That You Do: Many

educators are reluctant to reply to a question with the phrase "I'm sorry, but I do not know". For them, it is akin to an admission of stupidity. They are supposed to be the expert and if they do not have the answer they are often inclined to make one up. This often has lots of negative consequences that waste everyone's time and energy. You might pass along bad information or cause confusion that will last in your prospective student for a lifetime. Just because you have a BS degree, don't BS your way around a question. If you do not know an answer to a question, admit it and take the opportunity to research it. If you need to, tell the would-be student that you will get back to them with the answer later. If you lie your way through a question, the student will generally know and you will loose their respect.

239. Use Hands-On Activities: For all the kinesthetic learners in your group try to include some hands-on activities for them to participate in. Projects where they get to actually see and touch real fossils, measure them and work with them are a great way to teach concepts and techniques. Demonstrate the proper use of a tool or technique and then let them practice it. Let your kids make their own fossils using plaster of paris, in order to teach the concepts of fossilization. Have your tour group walk the length of an Apatosaurus so they can visualize the size of the great sauropods. Have your kids play in a sandbox or erosion table to teach the concepts of rapid burial. There are numerous examples of hands-on projects that can be used to teach paleo.

240. Be Careful Who You Let Hold the Real T. Rex Tooth: Rule #240 is for all of you paleo-technicians who work with real fossils. Some people, whether they are co-workers, grade-school children or museum visitors are just not careful enough around important specimens. Either they do not have the

#241- Seconds before disaster struck!

maturity to handle real specimens safely or they do not have the patience and understanding to be careful. Many do not understand the fragility and some will not grasp the importance. Hands-on specimens are a great way to teach people about paleontology, but be extremely careful with them or else you may have a mess on your hands. Look over your audience and pick and choose who gets to hold important pieces and who does not. When assigning projects in the laboratory for fossil prep, carefully pick and choose who works on what, based upon their skill level and their patience.

241. Ask Your Students Lots of Questions to Keep Them Involved and Awake: One of the best ways to tell if your students are paying attention is to ask THEM lots of questions. Learning involves a lot of give and take. You do not want to simply lecture to them and assume they are getting it. Instead, ask them questions, keep them involved. If they are afraid you might call on them to answer a question they will pay more attention so as to not become embarrassed if they have the wrong answer. This forces your audience to focus on you and not become distracted. They will learn more and you will be less frustrated.

242. Make Your Students Repeat Important Points: This rule is particularly important when working with younger students. It might feel like brain-washing to a certain degree, but if they chant back exactly what you have just said it does two important things. One, it forces them to pay attention, enabling them to learn more. Two, it lets you, the educator

know that you haven't lost the audience. For me, I make my younglings repeat the names of dinosaurs, geologic time periods, concept names, etc. periodically. "Can you say *Pachycephalosaurus?*" If they can (or they at least try), I know they are still paying attention. If they can't, I know I need to get more animated and more enthusiastic because I have lost them somewhere. Of course this technique does not work with older groups, as it will be perceived as condescending. Plus, after a certain age, most adults are incapable of saying "*Pachycephalosaurus*"!

243. Teach Both Sides of an Argument and Let the Students Make Up Their Own Minds: Your job, despite what many believe, is not to brainwash your students. Most issues in paleontology have more than one position. As a teacher, your job is to teach both sides of an issue. If you do not personally adhere to a particular hypothesis, then learn how to play devil's advocate and argue from the opposite side. For example, if you are reviewing the process of extinction, teach both the gradualistic AND the catastrophic arguments. Try very hard not to argue for either one or interject your personal opinion unless your students ask. If you only teach from one side of an argument then you are short-cutting your students and essentially brain-washing them. Too often professors interject their own ideas into the classroom. From this position of authority, especially when there is a grade involved, students may be more hesitant to disagree with the ideas of a professor. Putting your students in this position is unfair and leads to dangerous consequences. There is a time and place to be an advocate for a certain principle. The paleontology conference may be that place, but the university lecture hall is not. Present all valid hypotheses and let your students make up their own

minds. Your job is to teach; NOT to preach. Take off your researcher's hat and put on your teacher's hat.

244. Do Not Re-Write History: Rule #243

and #244 go hand in hand. Many professors, particularly those with a political axe to grind, interject their own opinions into their lectures. This is more of a problem in history or literature classes, but it does crop up from time to time in the paleo lecture hall. Professors need to be careful in how they present information. The conscious or unconscious omission of certain data may subtly or not so subtly lead students in a particular direction along the same lines of the various agendas of the professor. You may not like the methods employed by Cope and prefer the methods of Marsh, but omitting the contributions made by one without acknowledging the contributions made by the other is unfair and biased. You may not like the fact or wish to admit the fact that Marsh occasionally ordered his field workers to literally destroy certain specimens so Cope's men could not get access to them, but it happened and it needs to be discussed. You may not personally appreciate the role of commercial paleontology in the field, but omitting the positive things it has achieved is biased and short-sighted. You may feel Lamarkian evolution is invalid and no longer relevant, but failure to discuss it in its historical context is akin to giving students only part of the equation. You may dislike a particular technique, but not teaching it or at least, mentioning it, is not giving your preparators the full story and a full complement of ammunition. There will be plenty of time for providing your opinions and preferences. Do not re-write history by ignoring certain aspects you personally find unfavorable or distasteful.

245. Do Not Dumb Down to the Lowest Common Denominator: As a

teacher, your overall goal is to teach. You are not a baby sitter. You are not a social worker. Do not act like either. This applies from kindergarten through college and beyond. Far too many teachers approach education with the idea that they need to coddle and baby sit their students, dumbing down their lessons so that the class clown with the IQ of 65 at the back of the room, can get his D- and move onto the next poor slob. Your job is to teach and your student's job is to learn! If you dumb down your material you may keep a few afloat, but you wind up holding many more back. You need to challenge your students and force them to meet and exceed even their own expectations. You will find that as you raise the skill level in your lessons, you will raise the abilities of your students. When dinosaurs and ancient monsters are your subject, your students will be so eager to learn, they won't realize that they technically, "shouldn't be able to follow the concepts and thinking", you are presenting for a few more years. I am always surprised at how fast students pick up on supposedly advanced principles. First, teach a solid foundation, then, take them further than even you think is possible.

246. Keep Yourself Current: As stated in

several of the rules, paleontology is a rapidly changing and evolving science. As a teacher, you will need to give your students the most accurate, up-to-date information you possibly can. They will need that information. So, you will need to keep yourself as current and up-to-date as possible. Read and study frequently. Add Google news alerts for paleontology to your e-mail server. Once per month run a search on paleontological news groups. Investigate other related sciences, like astronomy, physics, biology and chemistry, to make yourself aware of changes in those disciplines. Just because you are now a teacher doesn't mean that you are no longer a student. Learning is a lifelong task.

VII. A FEW FINAL THOUGHTS FOR STUDENTS

247. Volunteer: One of the best ways to get your proverbial foot-in-the-door is to volunteer your time at the local museum. Most museums need volunteers as either preparators or as docents giving tours. There are thousands of specimens that need to be prepared and museum budgets often do not allow much in the way of paid staff. By volunteering your time and your skills, you will not only learn on the job many important paleontological duties, but you may also get the chance to meet and greet key members of the paleontological community. You never know how one of those connections might help you in the future. There are plenty of examples of people who volunteered for the prep lab only to be in the right spot at the right time, when a paying position opened up. If you do a good job, it may help you to get into the right school or it might pad your resume with skills or quality references. If you want to get into paleo you definitely want to explore volunteer options.

248. Undergraduates-Biological Route or Geological Route? There are two main paths to the kingdom of paleontology in college: the biological route and the geological route. Both routes are difficult and fraught with challenges. Both routes will teach you useful tools and skills you will need in your career. You will need to know a great deal about both disciplines in order to really master paleontology's greatest mysteries. Unfortunately, at most schools, you will have to choose one path or the other in which to concentrate.

The biological route involves heavy doses of anatomy, taxonomy and physiology and will concentrate on biological processes. You will focus your attention on ancient life in terms of living organisms. You will be trained to see the fossils before you as living animals, not as rocks. You will be able to imagine the soft tissue, consider the genetics and interpret form from function. You will study how one organism is related to another and focus on the finer points of evolution. The geologic route will involve heavy doses of sedimentology and stratigraphy, earth history and physical processes. You will learn about the rocks the fossils are encased in and will be able to interpret how they were laid down and in what paleo-environment. You will be trained to think long term and not short. You will learn about extinction and global processes. You will be trained to understand the geochemistry, the stratigraphy or the taphonomy of a fossil site.

Both paths are incredibly useful. Neither route is better than the other. Often a person who is planning on becoming a paleontologist can minor in one or the other, and this will help round out your education. Unfortunately, most, however, will get heavy doses of one discipline and only cursory doses of the other, because most people seldom have the time to double major or focus on both in sufficient detail. Irregardless, the best paleontologists are ones who have somehow, managed to be well trained in both.

You can tell quite easily which primary path a paleontologist took in their education by the style of their writing and the things they focus on with their research. Biologically trained paleontologists tend to focus more on the animal itself, whereas, geologically trained paleontologists tend to focus more on the dig site or the position of the animal in a stratigraphic context. Biologically trained paleontologists tend to focus on the finer details and smaller-scale, more direct implications. Geologically trained paleontologists tend to focus on the broader or larger picture implications. Biologically trained paleontologists tend to go into the more theoretical aspects of paleo, whereas geologically trained paleontologists tend to go more for the practical aspects of paleo. Biologists tend toward research and lab work, geologists tend towards exploration and fieldwork. These are of course not cut and dry and there are plenty of examples in contrast to the above generalizations.

Which path you decide to start out upon will depend on which direction you want to go in paleontology. Do you want to focus more on the practical or theoretical? Do you want to focus more on the taxonomy or the taphonomy? Are you more geared to research and analysis or fieldwork and observation? Before you choose a school and pick a path, you must answer all of these questions, looking deep within yourself for the answers.

249. The Squeaky Wheel Gets the Grease: In college, the more you interact with your professors and advisors the better. I'm not saying you should nag them day and night or stalk them in the cafeteria, but make sure they know who you are and what you want to do with your education. You do not want to be just another face in the department. You do not want to blend into the crowd. If you have questions, ask them. If you need advice, go and get it. Do not expect your advisors to always come and seek you out. You must be the master of your own destiny. The more they grow to know and hopefully, like you, the more opportunities and projects may come your way. If your advisor knows you want to go into paleontology, they will often drop any new and exciting projects, volunteer opportunities and independent research right into your lap. They won't know to do this if you quietly and meekly sit in the back of your classes without saying anything.

250. Get Lots of Field Experience: Not

only, should you volunteer in the lab, but you should also volunteer for field projects. Paleontology is very much a field science. The more opportunities you get to work on dig sites the better off your skills will be. The more geology you see, the more diverse your observational palette will become. Try to go on as many field trips as you can, even if those field trips are not directly related to paleontology. Invertebrate collecting trips, structural, sedimentological and general geology trips all have use.

251. Choose a Graduate School

Wisely: There are very few graduate schools out there offering advanced degrees in paleontology. In the United States, there are perhaps 20-30 that offer a Masters or Doctoral program. Your choices are limited, so make your decisions wisely. Make sure you personally visit each school to make sure you and it fit nicely together. Get to know the staff, their research interests, their methods, their connections and their eccentricities. Learn what they expect from you (teaching, research, etc.), should they accept you as a student and you accept them as a school. Discuss your research goals with your advisor. Make sure that the two of you are both on the same page.

252. Get Your PhD Whenever

Possible: As stated before, there are few paying positions in the world of paleontology. Most paying positions will require the minimum of a Master's degree to even be considered. There are strong preferences for those who hold a PhD. If you want to reach the levels of the most well- respected, then you will want to get your

#252- Thinking back to his preliminary college interview, Bob remembered the strange, shaking tic and the bizarre "whooping" yell, his new advisor had. Now... six months later, their cause seemed obvious.

doctorate. Your degree could be in paleontology specifically, biology, geology, taxonomy, museum science or any number of related things. You and I know that it's just a little piece of paper proving you have mastered the tests and challenges of school, but that degree will open lots of doors for you. Don't take a break from school and work for a few years. The more time you spend outside of school, the greater the chances are that you will be pulled in other directions. There are a few examples of paleontologists who stopped at a Bachelors or Masters degree and somehow willed themselves into a good position. A few more eventually did go back to school after many years of work and received their golden ticket. Those who have done things in this manner, however, are on a very short list.

253. When Opportunity Knocks, Open the Door: Opportunity is a strange creature. When you want it or need it the most, it seldom visits. When you don't need it, it camps outside your door roasting marshmallows in the hall. Paleontologists straight out of school, just starting out, need to be able to recognize a good opportunity when one presents itself and take full advantage of it. Opportunities are few and far in paleontology, so even if it doesn't seem to get you exactly where you want to go, any step in the right direction is a good first step. You will find that in paleontology the first step is always the hardest. If you see something close, take it, use it, learn from the experience and if you like, move on to the next one a few years down the road. The more experience you gain and the more connections you build, the more opportunities will come your way. You can afford to be picky later on in your life. Now is the time to grab the best option at the time, give it 110% effort and learn from it.

254. Choose Your Path Carefully: In

paleontology, there are a few main paths which you can take. Some will lead you to greatness and others to mediocrity, depending upon your skills, your determination and other pitfalls and curveballs along the way. Rule #248 talked about whether you should go along the biological path or the geological one. Rule #252 and #253 talked briefly about the degree level and direction your education should take. Rule #254 brings up another important path decision that will eventually come your way. Should your paleontological pursuits be more academic or more commercial?

There are three main groups of people involved in paleontology: academic, commercial and amateur. The primary differences between the three groups are how they are funded and the main purposes of their work. Academic groups make the bulk of their income from public funds. They receive wages from public sources through grants, stipends and museum or university paychecks. Their work is primarily, as the name implies, academic or solely for the pursuit of knowledge (and a nice paycheck). Commercial paleontologists receive the bulk of their funding from private sources, such as private companies or museums. Their income is derived through original fossil sales, cast specimen sales, tourism, resource assessment surveys for construction projects, or contract fossil preparation. Their work is primarily, as the name implies, for a commercial purpose (and secondarily as a pursuit of knowledge). Amateurs do their work without any pay whatsoever. They either volunteer for academic or commercial groups or work by themselves simply for the fun and enjoyment of it. The name amateur is really a misnomer as many so-called amateurs are really quite experienced and knowledgeable.

While there are some key differences, there are also many similarities between the groups. There are field workers, laboratory workers, researchers, technicians and educators in EACH of the three groups. Each group uses many of the same techniques and skills in their work as the others. There are often individuals who can not be pigeonholed into one of the three groups, preferring to straddle the brackets. Some successfully bounce back and forth. It is safe to say, that they ALL universally love the natural world and want to protect it, in their own ways. They simply disagree on the best methods and strategies for achieving this.

Most members of the groups get along just fine with one another and work on many joint projects for the sake of pure academic paleontology. For the last several decades however, some commercial and academic workers have unfortunately often been at odds with one another. The crux of the debate stems over fierce competition for fossil collecting sites. Even thought there are plenty of fossils and dig sites to go around, disagreements over how the best sites should be managed and handled cause a lot of turf wars. This has led to a wide variety of arguments and accusations from both sides. Some academics argue that commercial groups do not collect the necessary contextual data and what little data they do collect is tainted by a commercial motivation. They believe that commercial paleontologists are often no better than the average vandal or collector hack, which rip fossils out of the ground and then try to sell them on the often quoted, "black market". They fear that specimens recovered by commercial groups may wind up in the hands of private individuals where access to the specimens for research is limited or not available. Commercial paleontologists, on the other hand, argue that many fossil specimens would erode away to dust without their efforts and that responsible groups sell the majority of their finds to public museums anyway. They argue that the majority of specimens held in public museums were first discovered and collected by commercial paleontologists, amateurs or private landowners. They argue that fossils do have an intrinsic commercial value and that private landowners should be compensated for the specimens found on their land. They argue that many academics do a terrible job utilizing the lands that they do have control over. They argue that paleontology is something that everyone can and should enjoy and that science should not be limited to a few people in elitist ivory towers. They argue that it is possible and perfectly legitimate to make a living AND accurately collect the necessary data. Entire books could be written about the debate and turmoil caused by this 50 to 60 year-long turf war.

Whether you want to be involved in this debate or not, as a student of paleontology, you will unfortunately find yourself having to choose a direction and take a side. If you choose to go to work for a commercial group, even if they are a responsible, professional and caring one, expect to get some grief from certain members of academia. You may find yourself being blacklisted by some. You may find it difficult to get your research published. You may experience some stereotypical discrimination. I do not tell you this to try and steer you away from commercial paleontology. For many, it is the best and often only option. I tell you this, so that you are aware of the bias (from some) and can go onto this path with eyes wide open. For example, I have personally been doing commercial paleontology for over a decade and I am proud of it and my work. I do not rip bones out of the ground and I do not carelessly damage fossils or scientific data. I do not sell scientifically significant specimens to just any one off the street with a personal check. I do collect the scientific data and I do try to report that data to interested parties. Despite this, in my travels, I have faced a good deal of blind prejudice from certain members of the far-left, liberal, academic establishment. One in particular stands out.

At a Society of Vertebrate Paleontology meeting in Montana a few years back, I found myself having a conversation with a nice older woman during the nightly social event. We were both staring through the preparation windows of the museum lab, where the meet and greet was being held, and we began an otherwise pleasant and friendly discussion about the specimens on the other side of the glass. We discussed fossil preparation techniques and as I recall, duckbill dinosaurs. She seemed nice, perfectly rational, interesting and interested in my opinions. I enjoyed speaking with her and she seemed to be genuinely enjoying speaking to me. After about five or ten minutes of our pleasant conversation, the nice older woman glanced at my name tag and asked me what university I worked for. I replied that I did not work for a university, but was currently employed by a private company called *Triebold Paleontology*. The woman immediately turned white and was aghast. She asked, "is that (gasp)… is that a commercial company?", and I said, "Yes, it was." She looked as if she had just discovered she had been having a tea party with Hitler himself! She drew backwards (literally), shook her head and yelled at me in defiance and horror, "How could you do that!". She then turned and puffed away, disappearing into the surrounding crowds, with her nose in the air and shoulders pushed backwards. I was left standing there with my jaw dropped wondering what had just happened. In the span of less than 10 seconds, I went from being the nice guy who knew enough about fossil preparation to have a pleasant conversation with her, to a blood-thirsty, money-hungry, mouth drooling daemon out to destroy the fossils and defile the science. Had she produced a silver cross and hissed at me, I would have not been surprised (in hindsight). Not only was this rude and idiotic, and a vicious assault upon my character (without any real knowledge or evidence), but it shows the degree to which some members of academia literally DESPISE commercial paleontologists (without even knowing them). I have never encountered this woman before or since, nor do I ever care to. The point is, that irregardless of your attention to detail and commitment to science, if you find yourself involved in commercial paleontology expect some repercussions for your decision.

255. Do, or Do Not, There is no Try:

As out little green friend Yoda once said in the *Star Wars Movies*, "*Do or do not there is no try*". This is perhaps the best advice a young student can hear because it imparts a seriousness that they will need to have in order to pursue paleontology as a career. One needs to give the discipline their full attention in order to be successful. You do not "try" to learn it, either you do or you don't. Either you have the patience or you don't. Either you have the determination or you don't. You either put in the exceedingly long hours and make the difficult sacrifices or you don't. If you are able to do those things (following at least some of the advice contained in this book) and give it 110% effort, then you have a good shot at slowly working your way into this career. If you don't... have fun pushing papers at some insurance agency cubicle. There is no "try".

So... now that you've read the *Top 256 Rules of Paleontology*, take your passion, determination, dedication, courage, confidence, patience, wisdom and intelligence, give it your best, 110% effort and go out there and do it! Best wishes on your journey!

Oh, and finally, before I forget................

256. Try Not to Glue Yourself to the Bone!

VIII.RULES FROM THE REAL EXPERTS...

Ken Carpenter- *"Six things all students need to understand-*

1) Don't worry if you break it, that's what super glue is for

2) To be a good paleontologist, it has to be in your blood. It has to be a consuming passion

3) Don't travel with lawyers they are T. rex bait

4) No matter when you decide to stop digging, the skull will always lie an inch further

5) Don't pick on Marsh or Cope, they did the best they could with what they had- which wasn't much

6) Study an obscure group no one else is studying, that way you are sure to be the world's expert!"

Phil Currie- *"Rule #999: Never say "oops!" in a dinosaur quarry. Corollary to rule #999; Never keep quiet if you screw up!- All of us accidentally damage fossils from time to time and recognize that someone with less experience is going to make mistakes. It is all part of the learning! If you suddenly realize that the ironstone nodule you were smashing with your hammer is actually a bone disguised as a rock, stop and call in an expert! As long as more experienced people in the quarry know that a fossil has been damaged, steps can be taken to either fix (or at least control) the damage. I remember one amazing case where someone was trenching around an ornithomimid skeleton to provide drainage for a rainstorm. Several days later, we realized that the trench went straight through the end of the tail. Even though the tail was less than half an inch thick and tons of rock had been throw over the cliff from the quarry, a concerted effort to find the chunk of rock with the end of the tail proved successful!"*

Kraig Derstler- *"Study hard, then live and work with gusto-* It goes far beyond the classroom. *They should learn to ride a horse and drive a truck, live off the land and order Chateau Brianne, locate deeds in a courthouse and follow a cow trail, build a house and make a crate, set a broken bone and drink beer (in moderation, of course), work hard without complaining, lead, follow, quote Shakespeare and Spiderman, fly an airplane, wait out a three-day storm and quickly dodge an angry badger, win a hand of poker and smile when someone says no, write poetry and scribble an equally good contract, use a jackhammer and a pocketknife, interpret satellite photos and find M31. Literally, I have used every bit of background and experience in the course of my career as a vertebrate paleontologist."*

Mike Everhardt- *"Pay lots of attention to the small stuff-*
While it is often fun to pick up a fossil and be able to identify it, we often overlook the fact that it once was part (or all) of a living creature, and that there may be a lot more information associated with it than just it's identity. Far too many of the specimens collected during the early years of paleontology ended up being just objects in the storage area of a museum. Usually hurriedly pried out of the ground with whatever tools (picks and shovels) were available, these fossils were wrapped in newspaper and placed in a canvas bag until they were packed for shipment "back East." Many of them arrived battered and broken, and without much information.

Literally thousands of Late Cretaceous marine fossils from western Kansas that were collected by Marsh and Cope or their hired workers during the 1870s are sitting in collections today with minimal contextual information. You don't have to look very far to labels that indicate that the specimen was found in "Kansas" or "along the Smoky Hill River" or "southeast of Fort Wallace." Granted that there were no USGS maps or GPS available at the time, and that the early collectors knew little or nothing about the local geology, but I can also point out specimens collected by Prof. B.F. Mudge in the early 1870s that provide stratigraphic information and a legal description down to the quarter section. That kind of information allows modern workers to go to the locality and generate a lot more contextual data than just somewhere in "western Kansas." Apparently, neither Marsh nor Cope (and others) were particularly interested in that kind of information in their competition to name more and bigger species, and after a while, Mudge, Sternberg, Williston, and others simply stopped taking the time to record it.

Today we know much more about stratigraphy, and have GPS and detailed maps to enable to locate specimens very accurately. There is no excuse for not recording accurate information about the providence of the specimen. Beyond that, however, we also need to take the time to carefully examine the site and see what other information is associated with the remains. Early collectors, such as Mudge, Sternberg and Williston, were excellent collectors and were intimately aware of the condition of the specimens they were collecting. All of them have anecdotal stories about bite marks, broken and healed bones, soft tissue preservation, gut contents, and other valuable taphonomic information. Unfortunately, although they wrote about it occasionally, very little of that information is associated with the actual specimens where it was observed. Over the years that situation has improved significantly along with collecting techniques, and consequently we know more and more about the creatures that these fossils represent. In addition, as technology improves and we gain access to new preparation and preservation methods, more and better imaging processes, computer enhancement, and even biochemical analysis, these fossils are giving up more and more information.

The process of collecting the fossil, however, is where it all starts. You only get one chance to collect the specimen, one chance to do it right. Take your time and think it over BEFORE you start digging something out of the ground. Plan your dig. Decide what the best way is to collect the specimen. Assume that each specimen is unique and may contain new information, or even be a new species. Don' get in a hurry. The remains have been there, safe and sound, for millions of years; a little more time won't matter. Consider how much overburden needs to be removed to provide enough of a working surface. Measure, draw and photograph the dig extensively as you work on it (digital photographs are cheap... unless you don't have them). Do as little field preparation as possible; there will be plenty of time for that later. Record what you observe in the process; the condition of the remains, what kinds of things were they associated with, what was there and what wasn't there, all the little details that you think you will remember, but then have reason to question years later. Above all, label everything that you collect with field numbers and cross-reference them with a descriptive list in your field book. Don't find yourself unpacking the fossil later and wondering "Where heck did THAT come from?" Collect and label a sample of the matrix. This can be useful for more accurately dating the specimen, especially the remains of marine creatures. Be careful; be thorough. The job isn't finished until the paperwork is done.

Taking the time to do it right will save you lots of regrets in the future, and will provide much more useful scientific information in the process. Pay attention to the small stuff; it will pay off in a big way.

Don Glut- *"What better learning tool is there for kids than paleontology?-* Dinosaurs are particularly attractive to children, being often gigantic, always fantastic looking creatures, real monsters that actually lived, but because they're extinct (except for birds), they can inflict no harm. Paleontology can stimulate a child's imagination, wondering what extinct animals and plants were like when alive. They also have the power to inspire a child to read. At a very young age, the first real books I ever read were about dinosaurs. Through them, I was introduced to the field of paleontology and also science in general. Indeed the first author's names I learned were those of the paleontologists who had written those books. And if you want to bring up improving a youngster's spelling skills, I'd wager there are kids who can spell a 13 letter word like Tyrannosaurus while still having trouble spelling their own name. Yes, a lot can be said about paleontology beyond the study of fossils."*

Phillip Manning- *"To make a living in palaeontology, a sound grounding in optimism with a liberal dose of patience and enthusiasm goes a long way- Regardless of education, it is impossible to make a student passionate about science, as this is almost a fundamental adaptation to success in this field. To make a career out of this subject, you have to live, eat, sleep and dream about it. A very* *lucky few get to live that dream. Just make sure the lab book or field diary records as much passion as it does data...and never forget that this is one science where stating and recording the bloody obvious goes a long way!"*

Bruce Rothschild- *"If bridge is according to Hoyle, then paleopathology is according to Doyle- Once you have eliminated the impossible, what remains, no matter how improbable, must be the correct answer."*

Chris Ott- *"Try not to stress out over the little things- Paleontology should be fun. It shouldn't be considered "work". After all, no matter how much stress you have, the dinosaurs are still going to be dead in the morning."*

Greg Erickson- *"For paleontologists in academia these days, the pen (or the computer keyboard) is mightier than the pick- As in other academic fields, "publish or perish", is the key to success in paleontology. Professors at research universities are expected to conduct cutting-edge research and disseminate their findings in professional venues. Pay raises, job promotions, and securing grant money all hinges on these activities. Professors spend a vast amount of time compiling and analyzing data- and writing. Even those academics who conduct high profile paleontological field programs spend the majority of their time on campus studying and documenting the previous years findings, submitting grants for funding the upcoming field work, preparing lectures and completing various administrative duties- all of which require writing skills. I never imagined my middle school English teacher could teach me anything that would help my career in paleontology. Arguably, she helped me more than anyone."*

Neal Larson-*"Anyone can become a great paleontologist without investing time in school, but at school you learn how to learn, how to write and how to formulate and relate ideas- Succeeding in school gives you respect. But, anyone interested in becoming a*

paleontologist for money will never make it. Do it for passion not for anything else. Learn what you can about geology, primarily deposition and strata. Learn the chemistry of the rocks, how they were deposited and what happened to them between then and now. Study biology, it is only through the understanding of recent organisms that we can understand the ancient ones. And above all, always keep an open mind. The best ideas in paleontology come from new angles and off-the-wall thoughts!"

Larry Martin- "Fossils are common. Paleontologists are rare- We constantly worry about the wrong resource! If we support our paleontologists, there will always be plenty of fossils. The value of any object is based on what kind of story you can tell about it. It is the duty of the paleontologist to extract the best story possible from each fossil even if in the end the fossil is destroyed. Too many important fossils are rendered worthless, because we refuse to take a chance on preparation or analysis that might enhance the story that we can tell from them. At a time when entire museums are at risk, we need to make our fossils as exciting as we can, and as interesting to as wide an audience as possible. The more different studies that can be applied to any single fossil, the more important that fossil becomes and the more important each of these studies becomes. We need to share the thrill of discovery with our colleagues and with as broad a segment of the public as possible. We need not be a circus, but we must at least be an entertainment."

Scott Sampson- *"It ain't just fossils!- Vertebrate Paleontology sits on the cusp of two great disciplines: geology and biology. Over the past few decades, the field has become increasingly interdisciplinary, engaging tools, techniques and expertise from such diverse fields as physics, engineering, geophysics and molecular biology, among many others. Ramifications of this intellectual expansion, augmented by recent interest in such topics as climate change and extinctions include:*

1) The need for a broad education. Don't worry about fossils as an undergraduate. Get out there and get the basics on ecology, evolution, thermodynamics, and earth science. There will be plenty of time to specialize later.

2) The need to work in teams. Few top-notch VPer's work solo anymore. For example, the best field projects now include not just fossil specialists, but invertebrate workers, paleobotanists and geologists. Volunteers also make up a key portion of the team in most cases. While finding and naming new animals is important, we need to think of the whole ecosystem rather than select groups of organisms.

3) The need to communicate science to non-scientists. There is a great need for scientists to communicate their findings, and the nature of science more generally to the general public. In this time of ecological crisis, it is imperative that children and adults learn the basics about the workings of natural systems. Vertebrate paleontologists have a great advantage in this area, since their chosen field has broad appeal and serves as an easy access point to numerous areas of science. Interviews, popular articles, lectures, books and museum exhibits are all excellent venues to get the word out. So learn how to communicate to a general audience and do it."

Paul Sereno- *"Paleontology is a Cinderella science- combining the best part of a half a dozen fields from biology to geology to art- If you decide to discover fossils as part of your life's work it becomes the ultimate exercise program. It is a tough field, in that you must pursue it with a vengeance to make it. But, the best and shortest way I can describe what I do as a paleontologist is to say that it is "Adventure with a purpose!"*

Michael Brett-Surman- *"Three questions you should never ask a paleontologist- 1) Why are they always buried so far from the highway? 2) What did they use for water? 3) How many undiscovered species are there?*

Mike Triebold- *"Seven things I know about field work for any paleo-techs- This should get you through your first field experience:*

1) Don't dig like a maniac when your boss gives you the pick and tells you to remove a rock ledge. Pace yourself and let the pick do the work and you can swing the pick all day. Flex your muscles and slam the pick into the rock and you will not only wear yourself out in a half an hour, but the next day you will be worthless.

2) Don't dig yourself into a hole. When you are perimetering a site, or even an individual bone, step back and look at what you are doing. You are going to need to dig all the way around the bone eventually anyway, so don't work with your nose in a hole, trying to trace directly into the hillside. Remove overburden from all the way around where you think the bone or bones are, then move down on it from above. If you hit more bone, great, you have already given yourself an edge and it will go much, much faster.

3) Don't fixate on a single bone. This is under the category of "piddling around". If you uncover the end of a limb bone, move down where the shaft probably is and probe there, then move on. Don't sit in one spot with your brush and exacto knife, and expose any more bone than you need to. Amateurs are notoriously guilty of completely exposing beautiful bone in the field, only to have it ruined or damaged because they just didn't have the discipline to stop, perimeter it, jacket it and remove it, and do all the fun stuff back in the lab under controlled conditions. In my book, field prep is a no-no.

4) Drink lots of fluids. Drink more than you think you need. One of the primary causes of field inefficiencies is people who have headaches and weakness (and not thinking clearly) from not drinking enough water or other fluids. On a related note: wear sunglasses and a wide-brimmed hat if in the sun, along with long sleeves no matter how hot it is, and use the canopy at every opportunity.

5) *Keep your field camp cuisine simple. You will be exhausted and will not feel like making a fancy meal at the end of your day. Minimize perishables. Maximize foods that are compact and will last into the next millennium.*

6) *Obey your superiors. When the boss tells you to do something a certain way, there is a reason for it, usually borne of painful experience. You can discuss with him/her the reasons why at the end of the day when you are sitting in lawn chairs visiting. When you are working, do what you are told in the manner prescribed.*

7) *No whining! Yes, we know, there is no comfortable place to sit. It is too hot, too cold, too windy, too still, too many bugs, too wet, too dry etc ad infinitum. Buck up. You are there because of your passion for paleontology. Ignore the challenges to your convenience.*

Lawrence Witmer- *"To really understand the stories that fossils are trying to tell us, we need to look at the modern realm, animals living today to discover what the fossil clues really mean- The word Paleontology does not mean "the study of fossils", but rather literally translates to "the study of ancient being" or more loosely, the "the study of ancient life". How can we hope to comprehend ancient life without a rich understanding of modern life?"*

14338804R00129

Made in the USA
Charleston, SC
06 September 2012